Load Shape Development

Load Shape Development

Derek Schrock

PENNWELL PUBLISHING COMPANY
TULSA, OKLAHOMA

Copyright © 1997 by
PennWell Publishing Company
1421 South Sheridan Road/P.O. Box 1260
Tulsa, Oklahoma 74101

Library of Congress Cataloging-in-Publication Data

Schrock, Derek
  Load shape development / by Derek Schrock.
      p.   cm.
  Includes index.
  ISBN 0-87814-536-2
    1. Electric power plants—Loads—Statistical methods. 2. Electric
power consumption—Statistical methods. 3. Electric utilities—
Planning.   I. Title.
TK1191.S34   1997
621.31—dc21   97-2312
                                                                  CIP

All rights reserved. No part of this book may be
reproduced, stored in a retrieval system, or
transcribed in any form or by any means, electronic
or mechanical including photocopying and
recording, without the prior written permission
of the publisher.

Printed in the United States of America

1 2 3 4 5 03 02 01 00 99 98 97

Dedication

To My Parents

Luther J. Schrock
Janice E. Schrock

In Memory of Robert Norwood Rodgers

Contents

Figures and Tables		ix
Preface		xiii
Chapter 1	Load Shapes	1
Chapter 2	Terminology	11
Chapter 3	Sample Design	29
Chapter 4	End-Use Metering	53
Chapter 5	Data Collection	77
Chapter 6	Metered Data Analysis	101
Chapter 7	Engineering Analysis	129
Chapter 8	Statistical Analysis	167
Chapter 9	Transferring Load Shapes	201
Chapter 10	Beyond Load Shapes	221
Appendix	Assorted Data	239
Index		261

Figures and Tables

Chapter One
Figure 1-1 Typical residential oven usage pattern 2
Table 1-1 Common end-use categories by customer class 2
Figure 1-2 Load shape aggregation 3
Figure 1-3 Estimated lighting load shape 8

Chapter Two
Figure 2-1 Generic office load shape for a summer day 12
Table 2-1 Introductory load shape terminology 12
Table 2-2 Expanded load shape terminology 13
Table 2-3 End-use categories 15
Table 2-4 Two-digit SIC code classifications 16
Table 2-5 Residential customer classifications 18
Table 2-6a Daytype definitions for 16 daytypes 19
Table 2-6b Daytype definitions for 36 daytypes 20
Table 2-6c Daytype definitions for 48 daytypes 21

Chapter Three
Table 3-1 Distribution probabilities and their corresponding Z values 34
Table 3-2 Generic population 37
Table 3-3 Results for uniform stratification within building types 44
Table 3-4 Stratification of office buildings using the Delenius-Hodges method 47

Chapter Four

Figure 4-1	Demand variation over time	54
Figure 4-2	Time period to average	55
Figure 4-3	Moving average demand	55
Table 4-1	Typical metering capabilities and costs	58
Figure 4-4	Constant load with fixed operating hours	59
Figure 4-5	Constant load with variable hours	60
Figure 4-6	Variable load	61
Figure 4-7	Generic on-site data collection form for metering	63
Figure 4-8	Meter data collection from bakery	67
Figure 4-9	Metering plan for bakery circuit 2	69
Figure 4-10	Range checking of monitored data	75

Chapter Five

Figure 5-1	Potentially "complicated" heating questions	79
Figure 5-2	Less "complicated" heating questionnaire	80
Figure 5-3	Script questionnaire	83
Figure 5-4	Single business, single meter	89
Figure 5-5	Single meter, multiple businesses	90
Figure 5-6	Multiple meters, single business	91
Figure 5-7	Multiple meters, multiple businesses	92
Table 5-1	Important equipment characteristics	95
Table 5-2	General characteristics	97
Table 5-3	Typical square foot per ton values	97

Chapter Six

Figure 6-1	Single load research meter, single business	103
Figure 6-2	Single meter, multiple businesses	104
Figure 6-3	Multiple meters, single business	104
Figure 6-4	Multiple meters, multiple businesses	105
Figure 6-5	Single site, all end use metered	107
Figure 6-6	Single site, some end uses metered	108
Figure 6-7	Single site, some end uses combined	109
Figure 6-8	Single site, only whole-building metered	110
Table 6-1	Average loads for equipment being submetered	111
Table 6-2	Hourly load for five summer days	113
Figure 6-9	Average summer weekday example	114
Table 6-3	Average summer weekday end-use loads	116
Table 6-4	Grocery sector end-use loads (kW)	118
Table 6-5	Total loads for sites in grocery sector	120

Table 6-6	Grocery sector typical load end-use load shapes	122
Table 6-7	Typical whole-building load for grocery sector	124
Table 6-8	Average end-use load shapes	126
Table 6-9	Average whole-building load shapes	127

Chapter Seven

Figure 7-1	Box model lighting loads	133
Figure 7-2	Space model lighting loads	134
Figure 7-3	Occupancy sensor model	136
Figure 7-4	Daylight sensors	136
Table 7-1	Solar data for the 21st of each month	138
Table 7-2	Local standard time meridian (degrees)	139
Table 7-3	Coefficients for the solar variables	140
Figure 7-5	Industrial process box model	142
Figure 7-6	Industrial process on-off model	144
Figure 7-7	Industrial process sawtooth model	145
Figure 7-8	Industrial process sinusoidal model	146
Table 7-4	Linear hot water model results	147
Table 7-5	Square root hot water model results	147
Table 7-6	Age group hot water model results	148
Table 7-7	Hot water consumption model results	151
Figure 7-9	Cooking example 2 load shape	157
Table 7-8	Refrigeration and freezer annual energy use equations	158
Table 7-9	Refrigeration effect from cases	159
Table 7-10	Ratio of air handler CFM to motor hp	159
Figure 7-10	Design heating load schematic	160
Figure 7-11	Design cooling load schematic	161

Chapter Eight

Table 8-1	Annual energy use of residential homes	169
Table 8-2	End-use mix of residential homes	170
Table 8-3	Annual energy end-use regression coefficients	172
Table 8-4	Comparison of actual annual energy to estimated annual energy	173
Table 8-5	Monthly energy use	176
Table 8-6	Monthly heating and cooling degree days	178
Table 8-7	"Other" indicators by house	178
Table 8-8	Monthly regression results	179
Table 8-9	Regression results for house 1	180
Table 8-10	Regression results for house 3	181

Table 8-11	Regression results for house 10	182
Table 8-12	HVAC regression results for house 10	183
Table 8-13	Heating-only regression results for house 10	184
Table 8-14	End uses connected to load research meter	195
Figure 8-1	Comparison of engineering total and load research total	196
Table 8-15	Regression coefficients	197
Figure 8-2	Comparison of adjusted engineering total and load research total	197
Figure 8-3	Common HSEM findings	199

Chapter Nine

Table 9-1	Donor utility average grocery hourly loads	203
Table 9-2	Recipient utility hourly stratum and sector loads	204
Table 9-3	Donor utility stratum air-conditioning loads	205
Table 9-4	Donor utility average stratum air-conditioning loads	206
Table 9-5	Recipient utility stratum and sector air-conditioning loads	207
Table 9-6	Donor utility end-use loads for May peak weekday	211
Table 9-7	Donor and recipient utility factors	213
Table 9-8	Donor class-level end-use loads	214
Table 9-9	Donor and recipient residual loads	216
Table 9-10	Recipient utility class end-use and total loads	217
Figure 9-1	Donor and recipient total class loads	219

Chapter Ten

Table 10-1	Strata and population floorspace estimates	223
Table 10-2a	Annual consumption data for example 10-3	226
Table 10-2b	Strata and population end-use EUI estimates	227
Table 10-3	Strata and population demand intensities	231
Table 10-4	Estimated strata and population coincidence factors for air conditioning	233
Table 10-5a	End use data for example 10-7	235
Table 10-5b	Strata and population estimates of full load hours	237

Appendix

Table A-1	Summer hourly end use and load research data (kW) for example 8-4	240

Preface

As a utility consultant, I spent many years developing end-use load shapes for many clients and in the process analyzed load shapes for thousands of metered and unmetered facilities. I was the beneficiary of utility companies' policy of hiring outside firms to perform these tasks. Utility companies' preference to hire an outside firm to perform these tasks can be mainly attributed to the lack of in-house capabilities.

In early 1994, at the emergence of the deregulation process in the utility industry, I started wondering if it is possible to disseminate my knowledge of load shape development and applications in the new deregulated environment. I foresaw utility managers need to be well versed and competent in the process of load shape development.

This book is designed to provide guidance to a wide variety of practicing professionals and students. Some suggestions include the following:

- utility personnel who are planning to develop, or are in the process of developing, load shapes as well as other customer data;
- students and career up-starts who are learning about the utility industry;
- public utility commissions;
- utility consultants;

- power marketers and brokers; and
- finance professionals in the energy industry.

This book is not intended as a collection of the latest academic techniques and technologies which can be used to obtain load shapes; rather, it provides tried and true methods for developing load shapes in a **consistent, understandable,** and **straightforward** manner.

Also please note that the ideas in this book are biased by my experiences in developing load shapes. As data analysis is more of an art than science, my opinions on do's and don'ts when developing load shapes should not be viewed as absolute truths. Please consider these opinions as advice which you are free to disagree with as you see fit—if you have alternative opinions I would be interested in hearing them. Also, please treat this book as the starting point in the load shape development journey and let me know how your journey goes.

With the exceptions of Chapters 1 and 2, each chapter of the book is either dedicated to a unique step in the load shape development process or a unique analysis method to develop load shapes. Below is a brief summary of each chapter:

Chapter 1: Provides an introduction into what load shapes are and why they are needed by utilities.

Chapter 2: Provides information on common terminology used to describe load shapes as well as additional data which can be developed from load shape data. Additionally, end uses, building classifications, and end-use development methods are introduced and described.

Chapter 3: Provides an introduction to basic statistical sampling techniques. Statistical sampling can be used to leverage data from individual customers to other groups of customers. Basic sample design methods as well as basic statistical definitions are covered.

Chapter 4: Provides information on methods of metering to obtain end-use and whole-building metered data in buildings.

Chapter 5: Provides information on developing data collection surveys and performing on-site surveys to collect data to support end-use data development.

Chapter 6: Provides information on how to use end-use metered data to develop end-use load shapes.

Chapter 7: Provides information on how to use engineering models to develop end-use load shape estimates.

Chapter 8: Provides a broad array of statistical methods that can be used to develop end-use load shapes. Conditional demand analysis models, statistically adjusted engineering models, and a hybrid statistical engineering model are discussed.

Chapter 9: Provides information on how to transfer load shapes from one region of the country to another or from one year to the next.

Chapter 10: Provides post load shape analysis. After you have developed load shapes there are additional data that you can develop including full load hours and diversity factors.

I would like to take this opportunity to thank several people: first and foremost I would like to thank my editor, Jim Ferrier, for his continued support and patience during this process; a special thanks to John Stoops, Pacific Northwest National Laboratory, who helped me with book content and encouraged me to pursue the book; thanks to Susan Haselhorst, Andy Schon, and Bill Huss (all of whom I worked with at XENERGY, Inc.) for their teachings and support during the early stages of my career; and Dennis L. O'Neal and David E. Claridge for teaching me about the energy industry and energy use in buildings and employing me while at Texas A&M University. Finally, I would like to thank my parents for always believing in me.

Derek Schrock

1
Load Shapes

Introduction

First and foremost, load shapes provide a means of understanding how much energy is being used at different times of the day, week, season, or throughout a year. Put another way, load shapes help convey the patterns of energy use. Figure 1-1 provides a graphical depiction of the energy usage pattern (or load shape) for an oven in a typical residence.

The second important characteristic of load shapes is that they can represent an infinite range (or levels) of energy use patterns: from the small energy use of a single appliance or light bulb at a single facility, to the total energy being utilized at a single facility, all the way up to the energy being used by a large group of utility customers, perhaps even the energy usage patterns of every customer that the utility services. When the energy use patterns are being developed for groups of equipment with similar functions, the results are commonly referred to as *end-use* load shapes. An example of this would be to examine the energy use pattern for all of the refrigeration cases and compressors in a grocery store. Some of the most common end-use categories for the commercial, residential, and industrial *classes* are presented in Table 1-1.

Another common level of load shape data is the entire energy use pattern from a single electric meter, which, in a lot of cases, also represents the energy use pattern for a single facility or customer. When load shapes represent the energy use pattern for a single

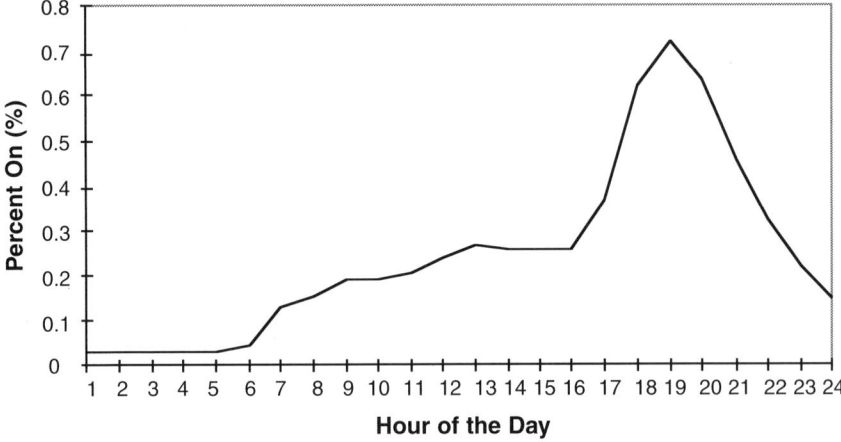

Figure 1-1. Typical residential oven usage pattern.

Table 1-1. Common end-use categories by customer class.

End-Use Category	Commercial	Residential	Industrial
Air conditioning	•	•	•
Space heating	•	•	•
Interior lighting	•	•	•
Miscellaneous equipment	•		•
Domestic hot water	•	•	•
Computers	•		
Cooking	•		
Refrigeration equipment	•		
Ventilation	•		•
Exterior lighting	•		•
Process equipment			•
Motors			•
Stoves/ovens/ranges		•	
Refrigerators/freezers		•	
Televisions/stereos		•	
Dishwasher		•	
Clothes washer/dryer		•	

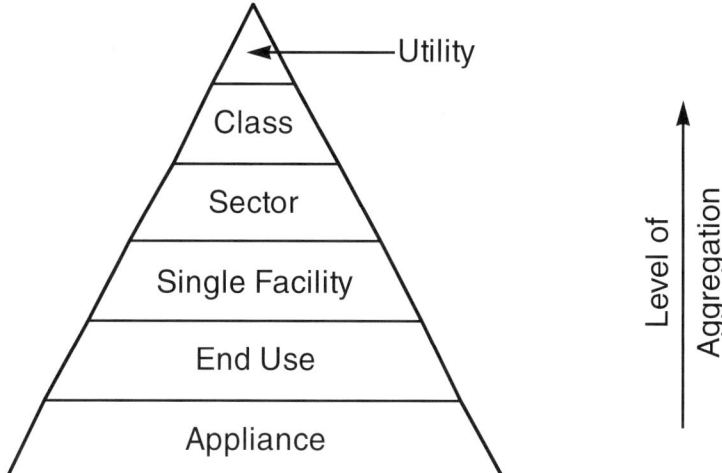

Figure 1-2. Load shape aggregation.

utility meter, they are commonly referred to as *whole-building* load shapes. Groups of customers which share similar functions are commonly grouped into categories, or building *Sectors*. Examples of commonly used sector categories include office buildings, restaurants, and hospitals. The next level of aggregation involves grouping sectors into even larger groups, or *Classes*, of customers. Most utilities have at least three customer classes: commercial customers, residential customers, and industrial customers. If a utility services rural areas, they may include an agricultural class. Figure 1-2 provides a graphical overview of the various levels of aggregation that load shapes are commonly applied to.

Why Are Load Shapes Important to Utilities?

Electrical utilities have historically made large use of end-use load shapes in the energy and load forecasting, integrated resource planning and demand-side management (DSM) evaluation arenas. More recently, natural gas utilities have begun to utilize end use load shapes for the same purposes. For energy and load forecasting, utilities are commonly required to submit a prediction to a public utility commission of what their future capacity needs will be considering such factors as: current base load, the expected change in the number of residential homes, commercial stores, and industrial facilities, and the change in equipment efficiencies over time. However, in order to obtain as accurate of an estimate as possible the current base load and change in equipment efficiencies needs to be quantified by end-use categories such as those shown in Table 1-1. Load shapes are useful in these applications since they can provide the existing energy use or demand for typical utility customers.

Based on the forecast supply and demand curves, a utility may wish to examine its current and future state of generating capacity. In the simplest sense, if a utility is capacity constrained during a particular period it will examine the end-use loads during that time to determine which of its customers' energy use patterns can be altered through a utility DSM program. Alternatively, if a utility has a large amount of excess capacity during a particular period, it may wish to inspect the available end-use marketing opportunities that are available at that time, such as a DSM marketing program to promote high-efficiency outdoor lighting. Again, end-use and whole-building load shapes quantify how energy is being used by a utility's customers and can provide a good starting point for the utility.

An increasing number of Public Utility Commissions (PUCs) require that utilities file an integrated resource plan, which examines both the supply and demand curves along with the options of building new generation purchasing power from other power producers, installing distributed utilities nearer on their customers' sites, or implementing DSM programs to alter the way their customers utilize energy. One of the primary goals of the integrated resource planning process is to determine the lowest cost method of meeting predicted future energy requirements.

When DSM programs are selected and implemented, it is because they have been determined to be cost effective based on assumed energy and/or demand savings and program administrative costs. In order to improve these assumptions, several PUCs require that evaluations be performed to determine (1) the true savings and (2) program costs, respectively known as impact evaluation and process evaluation. There are several methods for performing the impact evaluation: end-use metering, whole building load research data, billing history analysis, building simulation, and other statistical methods. Several of these approaches can be used to generate and support the development of end-use load shape impacts—which capture the change in energy use patterns resulting from the DSM program.

In addition to providing an overview of the current state of end-use load shape development needs by utilities, such as energy and load forecasting and integrated resource planning and evaluation, we need to guess at the potential future needs of utilities and how the move toward utility deregulation may alter how load shapes are utilized. Two areas that are expected to grow rapidly in the next few years are (1) distributed generation and (2) targeted DSM programs. One impact of both of these areas is that utilities need to better define and understand how energy is being used by their customers in order to be competitive. One example of this future may be that instead of a utility implementing a refrigeration efficiency program for all of the grocery stores in its service territory, it will develop DSM programs for targeted (e.g., smaller) groups of customers; perhaps the utility will only target those grocery stores that have electric heat and have more than 20 refrigeration cases.

Load Shape Development Steps

Now that you understand what a load shape is and some of the reasons why it is so critical to the utility industry, an overview of the entire load shape development process can be provided. Let us assume that a utility wants to develop end-use load shapes for each of its customer classes (e.g., office buildings, restaurants, hospitals, etc.). Several of the processes that have been used successfully by utilities to develop load shapes include one or more of the following steps:

- sample design
- perform on-site metering
- perform on-site data collection for selected customers
- develop end-use load shapes for each customer
- develop end-use load shapes for the population

The remainder of this section will be devoted to providing an introduction to each of these load shape development steps.

Sample Design

The only way to be 100 percent certain that the load shapes being developed provide an absolutely perfect understanding of the energy used by all of a utility's customers would be to analyze the load shapes for each utility customer. Unfortunately, this would be an enormously time-consuming and expensive proposition—one that neither the utility nor its customers want to pay for. Therefore, in order to reduce the cost of developed load shapes a utility will study the load shapes for a fraction, or *sample,* of its customers. But how do utilities know how many facilities to study to develop an accurate representation of their utility? The answer is that they perform a sample design.

The purpose of a sample design is to ensure that the results are representative of the population the sample is being selected from. Furthermore, the sample design allows the utility to study effects of sample size on statistical confidence. If the utility wants to increase the statistical confidence of the load shapes, it can increase the number of facilities (or *observations*) being analyzed. Conversely, if it doesn't care as much about the precision of the final load shapes, it can choose to study fewer facilities. Optimization methods are also available that can minimize the number of observations that need to be analyzed, such as Dalenius-Hodges stratification with Neyman allocation.

Metering Equipment and Facilities

Metering involves collecting information on the power or energy being used by an appliance, end use or an entire facility during normal operation. Some more common types of meters include spot meters, run-time meters, and data loggers. Spot meters use one or more *current transducers* to obtain an estimate of the instantaneous power being used by an electrical appliance, or circuit. An example of a spot meter application would be to determine the power used by a mainframe computer of a facility during normal operating hours. Run-time meters are used to collect the number of hours that an appliance or circuit is on (or operating) during a typical day or week. Run-time meters work best for energy using devices that have a relatively consistent pattern of usage throughout the day, week, and year, such as indoor lighting. Data loggers can be used to obtain the actual load shape, or energy pattern, for a piece of equipment, an entire circuit, or even an entire facility. Most utilities actually have many years' worth of data collected for the total energy use of facilities for a sample of their customers. This data is commonly referred to as *load research data* since utilities are required to collect it for research purposes. The meters that are used to collect the load research data are called *load research meters*. One nifty feature of data loggers is that they can generally store the energy use patterns of devices and electric circuits, or data points, for a long period of time (weeks or months). Furthermore, some data loggers can store information from several different data points simultaneously.

Even though actual metering provides a more accurate representation of how energy is being utilized at a site than most other methods, it still has its drawbacks. If you want to use whole-building load data, the biggest problems occur when the utility meters do not provide energy to the entire facility, or the facility has more than one utility meter but only one of them collects load research data. In this case you need to try to determine which devices inside the facility are obtaining energy from the load research meter. Another problem that is commonly found in retail shopping malls is that adjacent stores may share the same electric meter. In this case, you are probably best off surveying data in both stores to ensure that you understand what equipment is connected to the load research meter.

On-Site Data Collection

On-site data collection involves having energy auditors go to a customer's facility and perform a survey on the energy using equipment and characteristics of the site. Some of the other methods of collecting information for facilities include sending a mail survey for someone at the facility to fill out or calling the people on the phone. However, be forewarned that these two methods are inherently less accurate than having a qualified energy auditor perform a survey at the site. The most important types of information to collect while on-site are the types of equipment, the schedule of operation for the equipment, and the nameplate *capacity* of the equipment. The equipment capacity is a rating of the maximum power that the equipment would draw if it was running under a maximum load condition.

Develop Facility End-Use Load Shapes

Once the on-site data has been collected from the required facilities, estimates of the end-use load shapes can be developed for each facility. Methods that can be used to develop end-use load shapes include engineering estimates, building simulation models, and statistical methods such as conditional demand analysis. Engineering estimates can provide quick and simple estimates of the load shape patterns using equipment capacity and schedules. For example, if lights were known to come on at 7:00 A.M. and remain on until 6:00 P.M. during the week, were off for the remainder of the time, and the lights had a total connected load of 5 kilowatts (kW), the load shape shown in Figure 1-3 could be generated. While simple engineering estimates might work well for developing simple, consistent end-use load shapes, such as indoor lighting, more complicated end uses would require more sophisticated techniques. Probably the most complicated end uses encountered in facilities are the heating and cooling systems. The reason is that in order to predict the energy use, the outdoor weather conditions, how much sunshine is striking the building, the mass and thermal properties of the facility walls and roofs, the glass and thermal properties of the glass in the facility, not to mention the operational characteristics of all the things in the facility that produce heat (including people), the efficiencies and characteristics of the heating and cooling systems need to be known.

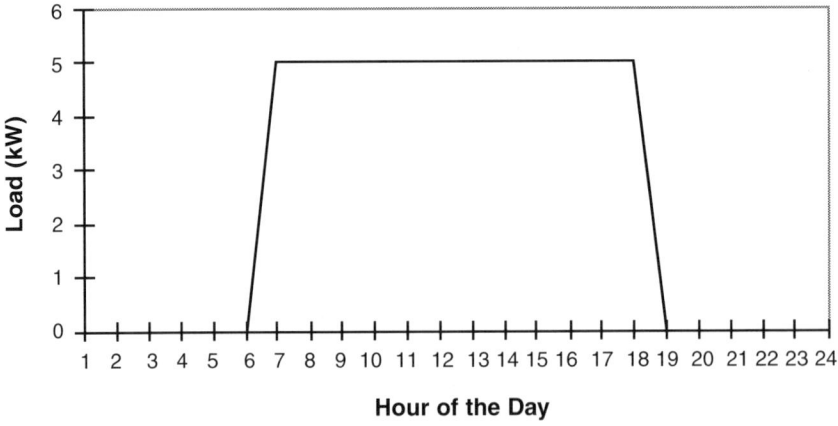

Figure 1-3. Estimated lighting load shape.

Furthermore, even if we could manually follow all of the equations required for this prediction, we would need to perform it for different times throughout the year. Fortunately, we have *building simulation* models that handle the really complicated issues and produce estimates of the energy use patterns for the heating and cooling systems.

Statistical methods, such as conditional demand analysis (CDA), provide a middle ground between simple engineering equations and building simulation models by using engineering parameters as surrogates (or priors) for end-use estimates. For example, a surrogate for the heating energy use might combine the difference between indoor and outdoor temperature, the U-value of the building shell, and the square footage area of the building shell into a simple heat loss estimation. The CDA method compares the total load at a given facility against surrogates for each of the end uses in the facility to provide correction factors for each end use. These hourly correction factors can then be applied to the end-use surrogate to produce a load shape for a typical day. By following one or more of the above methods, end-use load shapes can be developed for any facility.

Develop Population Load Shapes

As a final step, the results from the sample of customers need to be transformed into a load shape which presents the energy use for the population that they represent. The simplest way to perform this task is to apply a *population weight* to each of the customers in the sample. The population weight is the ratio of the number of utility customers divided by the number of customers in the sample for a given group of customers. It is important to note that the groupings of customers are the same ones developed in the sample design process. For example purposes, assume the following:

- from the sample design, it was determined that there were 150 grocery stores in the population of customers served by the utility
- the sample design also indicated that nine grocery stores needed to be surveyed to obtain a representative sample
- load shapes were developed for nine grocery stores

The population weight for each of the nine grocery stores in the sample would be estimated as 150 divided by nine or 16.7; implying that each of the grocery stores in the sample is representative of 16.7 grocery stores in the population of utility customers. The final step in applying the population weights is to sum the product of the demand values in the customers' load shapes by the population weight for each customer.

2
Terminology

Introduction

One of the biggest hurdles that the utility industry faces, whether it is performing metering activities, evaluation of its DSM programs, or load forecasting, is communicating information both internally in its company and externally to its customers. In the case of electric utilities, there has historically been a need to distribute information to other electric companies as well as public utility commissions or other agencies. With the emergence of real-time data collection and control and the information superhighway, data management and data consistency have never been as critical as they are becoming today. The terminology presented in this chapter is a combination of terms defined by others (Western Area Power Administration, 1990) as well as original definitions.

Load Shape Terminology

As a starting point, let's examine some of the common information that can be gathered from load shape information for use by utilities. The first introductory concepts are connected load, peak load, observed load, peak hour, and system peak hour as depicted in Figure 2-1, which is a representation of the total, or whole-building, power usage of a generic building on an hour-by-hour basis.

Given the above terminology descriptions, presented in Table 2-1, we can now begin to calculate information that utilities can use to

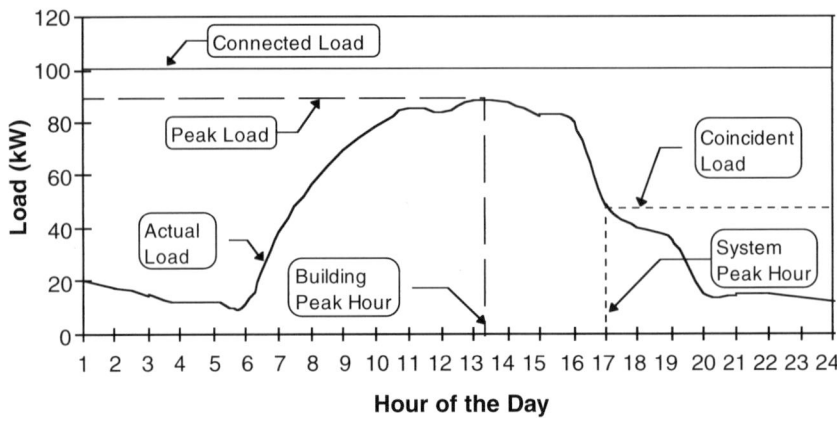

Figure 2-1. Generic office load shape for a summer day.

Table 2-1. Introductory load shape terminology.

Connected load	For a system, such as a building or electrical circuit, the connected load is the sum of the nameplate capacities of the energy consuming devices connected within the system.
Nameplate capacity	The nominal rated capacity for a given piece of equipment.
Peak load	The maximum load as calculated by the metering device for a given time period.
Actual load	The average load of the equipment at any period in time.
Peak hour	The hour of the day, for the day on which the peak load occurs.
System peak hour	The hour on which the electric utility system produces the largest quantity of power for the period in question. This value is commonly quantified on either a seasonal (e.g., summer and winter) or monthly basis.

Table 2-2. Expanded load shape terminology.

Diversity factor	Diversity refers to the fraction of connected load being utilized amongst a group of appliances or buildings during a given time period if they do not all peak at the same time.
Coincidence factor	The fraction of peak load that is being utilized during a given time period.
Part load factor	The fraction of rated load that an individual piece of equipment is operating at during a given time period.
Full load hours	The amount of time that a piece of equipment or end use would operate in a given time period if it were operating at peak load conditions. For electrical equipment, annual full load hours can be calculated by dividing the annual energy use for the equipment by the observed peak load of the equipment.
Market share	The percent of end-use load which utilizes a given fuel (electric, natural gas, propane, etc.).
Appliance saturation	The number of appliances which the average customer has in his/her residence (e.g., 1.2 freezers per home).

quantify how a building, end-use, or single appliance is being utilized at a single site or set of sites. Terms will now be provided for diversity factor, coincidence factor, and full load hours as shown in Table 2-2.

Given a consistent set of terminology that can be utilized at the appliance, end-use, building, sector, or utility system level, we can now develop simple engineering relationships between the various terms. Equations 2-1, 2-2, and 2-3 present engineering equations of relationships between terminology at different levels of data resolution. It is important to remember that for the appliance level of resolution, both the diversity factor and coincidence factor are unity; and for the building level of resolution, the coincidence factor is unity.

$$Actual\ Load = (Connected\ Load \cdot Part\ Load\ Factor \cdot Diversity\ Factor)$$

(Equation 2-1)

$$Diversity\ Factor = \frac{\sum Peak\ Load}{\sum (Connected\ Load \cdot Part\ Load\ Factor)}$$

(Equation 2-2)

$$Coincident\ Load = (Actual\ Load \cdot Coincidence\ Factor)$$

(Equation 2-3)

Energy and Demand Savings Estimates

Using the terminology presented in this chapter, simple engineering equations can be developed to estimate both energy savings and system peak demand savings that result from DSM programs. First, for the case where a site is installing more efficient equipment, but the hours of operation for the equipment are the same in both the pre- and post-installation periods, the energy savings is equal to the product of the change in connected load times the part load factor and the full load hours, as shown in Equation 2-4. Additionally, the demand savings is equal to the change in connected load times the part load factor, the diversity factor, and coincidence factor, as shown in Equation 2-5.

$$Energy\ Savings = \begin{pmatrix} \Delta\ Connected\ Load \cdot Part\ Load\ Factor \\ \cdot Full\ Load\ Hours \end{pmatrix}$$

(Equation 2-4)

$$Demand\ Savings = \begin{pmatrix} \Delta\ Connected\ Load \cdot Part\ Load\ Factor \\ \cdot Diversity Factor \cdot Coincidence\ Factor \end{pmatrix}$$

(Equation 2-5)

The second common case is where controls are added to the equipment to change, or control, the hours of operation. For this case, the demand savings are zero, if it assumed that there is no change in operation during the peak period. However, the energy savings can be calculated as the product of the change in full load hours times the connected load and part load factor as shown in Equation 2-6.

$$Energy\ Savings = \begin{pmatrix} \Delta\ Full\ Load\ Hours \bullet Part\ Load\ Factor \\ \bullet Connected\ Load \end{pmatrix}$$

(Equation 2-6)

End-Use Categories

One traditional method that has been successfully utilized by utilities when designing DSM programs is to target those end uses that have either a large energy consumption or contribute significantly to the utility's system peak demand. Typical end-use categories for commercial, industrial, and residential customers are presented in Table 2-3.

Table 2-3. End-use categories.

Commercial Sector	Industrial Sector	Residential Sector
Interior lighting	Interior lighting	Lighting
Exterior lighting	Exterior lighting	Refrigeration
Office equipment	Office equipment	Domestic hot water
Computer equipment	Computer equipment	Space heating
Domestic refrigeration	Domestic refrigeration	Ventilation equipment
Hot water heaters	Domestic water heat	Air conditioning
Space heating	Space heating	Cooking equipment
Ventilation equipment	Ventilation equipment	Cooktop/stoves
Air conditioning	Air conditioning	Microwave ovens
Cooking equipment	Cooking equipment	Televisions
	Motors	Water bed heaters
	Process equipment	Pool/spa pumps
	Process water heat	Pool/spa heaters
	Pumps	

Building Type Categories

In addition to defining end-use categories, utilities are always trying to better understand how energy use patterns vary among customers. One method that has been historically used to categorize customers in the commercial and industrial sectors is to separate them according to their primary business activity, SIC code. Table 2-4 presents the primary two-digit Standard Industrial Classification (SIC) code classifications. Table 2-5 presents classifications that have been used to separate residential customers.

Table 2-4. Two-digit SIC code classifications.

SIC	Classification	SIC	Classification
Agricultural, Forestry, and Fishing		*Manufacturing*	
01	Agricultural production—crops	20	Food and kindred products
02	Agricultural production—livestock	21	Tobacco
		22	Textile mill products
07	Agricultural services, etc.	23	Apparel and other textile products
08	Forestry	24	Lumber and wood products
09	Fishing, hunting, and trapping	25	Paper and allied products
		26	Printing and publishing
		27	Printing and publishing
Mining		28	Chemicals and allied products
10	Metal mining	29	Petroleum and coal products
11	Anthracite mining	30	Rubber and plastic products
12	Bituminous coal and lignite mining	31	Leather and leather products
		32	Stone, clay, and glass products
13	Oil and gas extraction	33	Primary metal industries
14	Mining/quarrying of nonmetallic minerals	34	Fabricated metal products
		35	Machinery, except electrical
		36	Electric and electronic equipment
Contract Construction			
15	Building construction—general contractors	37	Transportation equipment
		38	Instruments and related products
16	Construction—other		
17	Construction—special trade contractors	39	Miscellaneous manufacturing industries

Terminology

Transportation, Communications, and Electric, Gas, and Sanitary Services
40 Railroad transportation
41 Local and passenger transportation
42 Motor freight transportation warehousing
44 Water transportation
45 Transportation by air
46 Pipeline transportation
47 Transportation services
48 Communication
49 Electric, gas

Wholesale Trade
50 Durable goods
51 Nondurable goods

Retail Trade
52 Building materials, hardware, etc.
53 General merchandise
54 Food stores
55 Automotive dealers and gas stations
56 Apparel and accessories
57 Furniture, home furnishings, etc.
58 Eating and drinking places
59 Miscellaneous retail stores

Finance, Insurance, and Real Estate
60 Banking
61 Credit agencies other than banks
62 Security brokers dealers, etc.
63 Insurance carriers
64 Insurance agents and brokers
65 Real estate
66 Combination insurance, loan, law, etc.
67 Holding and other investment companies

Services
70 Hotels, motels, and trailer parks
72 Personal services
73 Business services
75 Automobile repair and service
76 Miscellaneous repair services
78 Motion pictures
79 Amusement and recreational services
80 Medical and other health services
81 Legal services
82 Education services
83 Social services
84 Museums, art galleries
86 Nonprofit membership organizations
88 Private households
89 Miscellaneous services

Public Administration
91 Executive, legislative except finance
92 Justice, public order, and safety
93 Public finance, taxation and policy
94 Admin. human resources
95 Admin. environmental quality/housing
96 Admin. of economic programs
97 National security and international affairs
99 Nonclassifiable establishments

Table 2-5. Residential customer classifications.

Residential Classifications
Single family
Duplexes
Multi-family
Manufactured homes

Daytypes

When utilities collect whole-building metered data from buildings over a years' time span, they typically store the data for every day of the year in 15-minute demand increments. However, when it is necessary to analyze these data and develop end-use load shape estimates, the amount of data can become unwieldy. Therefore, several software packages and database libraries collapse these data into daytype groupings. A daytype is a single day, comprised of 24 hourly values, which represents a larger number of actual days. Three of the most common daytype definition storage schemes are the 16-daytype, 36-daytype, and 48-daytype representations. In the 16-day representation, the year is represented by four seasons (summer, spring, winter, and fall) and four types of days within each season—peak weekday, average weekday, peak Saturday, and peak Sunday—for each season. In the 36-daytype scheme, the year is categorized by month with three daytypes for each month—peak weekday, average weekday, and average weekend (which is a combination of both Saturday and Sunday). The 48-daytype resolution expands on the 36-daytype scheme by defining four types of days for each month—peak weekday, average weekday, average Saturday, and average Sunday. The full description of the daytypes for each system are presented in Table 2-6a through 2-6c.

Table 2-6a. Daytype definitions for 16 daytypes.

Daytype ID	Description
1	Winter peak day
2	Winter average weekday
3	Winter average Saturday
4	Winter average Sunday
5	Spring peak day
6	Spring average weekday
7	Spring average Saturday
8	Spring average Sunday
9	Summer peak day
10	Summer average weekday
11	Summer average Saturday
12	Summer average Sunday
13	Fall peak day
14	Fall average weekday
15	Fall average Saturday
16	Fall average Sunday

Database Libraries

There are currently databases available to electric utilities which contain both generic whole-building and end-use load shapes for typical building types and specified daytypes. In addition to databases of typical load shape data, there are databases such as the Electric Power Research Institute's DSManager that contain generic impact load shapes that result from DSM programs. Impact load shapes represent the difference in the load shape from before DSM measures are installed to after they are implemented. If a utility has no whole-building or end-use load shapes already developed, these libraries can provide a quick and inexpensive source of data. However, it is important to realize that potential problems exist when using generic load shapes. First and foremost, you may not readily know whether the generic load shapes are representative of another utilities' population or are developed by examining a prototypical building.

Table 2-6b. Daytype definitions for 36 daytypes.

Daytype ID	Description
1	January peak day
2	January average weekday
3	January average weekend
4	February peak day
5	February average weekday
6	February average weekend
7	March peak day
8	March average weekday
9	March average weekend
10	April peak day
11	April average weekday
12	April average weekend
13	May peak day
14	May average weekday
15	May average weekend
16	June peak day
17	June average weekday
18	June average weekend
19	July peak day
20	July average weekday
21	July average weekend
22	August peak day
23	August average weekday
24	August average weekend
25	September peak day
26	September average weekday
27	September average weekend
28	October peak day
29	October average weekday
30	October average weekend
31	November peak day
32	November average weekday
33	November average weekend
34	December peak day
35	December average weekday
36	December average weekend

Table 2-6c. Daytype definitions for 48 daytypes.

Daytype ID	Description
1	January peak day
2	January average weekday
3	January average Saturday
4	January average Sunday
5	February peak day
6	February average weekday
7	February average Saturday
8	February average Sunday
9	March peak day
10	March average weekday
11	March average Saturday
12	March average Sunday
13	April peak day
14	April average weekday
15	April average Saturday
16	April average Sunday
17	May peak day
18	May average weekday
19	May average Saturday
20	May average Sunday
21	June peak day
22	June average weekday
23	June average Saturday
24	June average Sunday
25	July peak day
26	July average weekday
27	July average Saturday
28	July average Sunday
29	August peak day
30	August average weekday
31	August average Saturday
32	August average Sunday
33	September peak day
34	September average weekday
35	September average Saturday
36	September average Sunday

(cont'd)

Table 2-6c. Daytype definitions for 48 daytypes *(cont'd)*.

Daytype ID	Description
37	October peak day
38	October average weekday
39	October average Saturday
40	October average Sunday
41	November peak day
42	November average weekday
43	November average Saturday
44	November average Sunday
45	December peak day
46	December average weekday
47	December average Saturday
48	December average Sunday

The major difference between representative and prototypical load shapes is that representative load shapes are built up from an aggregate, or mix, of utility customers that utilize a range of equipment types (e.g., the air conditioning load shape may represent 40 percent rooftop DX units, 30 percent central DX units, and 30 percent air-to-air heat pumps) and operating schedules, while a prototypical building may only represent a single set of equipment types (e.g., the air conditioning system may only represent a centralized chiller) and one operating schedule.

The second problem is that you may not have access to the market research data or weather data that corresponds to the generic load shapes. This is important if you want to try to transfer the load shapes from the generic region to your utility.

Load Shape Development Methods

Some of the most common end-use load shape development methods currently used by utilities include end-use metering, engineering models, statistical models, and statistical/engineering models. This section will introduce each of these methods as well as provide an overview of typical steps involved in developing end-use load shapes using each of the methods.

End-Use Metering

End-use meters can be installed at the appliance or electrical circuit levels of buildings to record how groups, or individual pieces, of equipment utilize energy either instantaneously or over a longer period of time. Several types of metering equipment are commonly used to gather energy information including: instantaneous spot meters, run-time meters, and data logger equipment. The steps required to properly perform an end-use metering study for a group of facilities are:

Step 1: Sample design.

Step 2: Develop metering plans for selected sites.

Step 3: Check and calibrate meters.

Step 4: Install metering equipment.

Step 5: Perform quality control checks in the field.

Step 6: Perform quality control on recorded data.

Step 7: Aggregate end-use data to the desired level.

Steps 1 and 7 are only required if you are estimating end-use load shapes for a population from a sample of the populations' facilities.

Engineering Models

While end-use metering develops end-use load shapes from actual data recorded in the field, engineering models approximate end-use load shapes by assuming physical attributes for the equipment in facilities, such as the power consumed by a piece of equipment it is running at part load or full load, as well as operational characteristics of the equipment, including when the equipment is drawing power. The data required for these physical and operational assumptions may be developed through several methods, including on-site data collected at facilities, mail surveys filled out by facility personnel, or the engineering judgment of the person developing the load shape data. These data collection methods are further explained later in this book. The steps required to develop end-use load shapes using engineering models include:

Step 1: Sample design.

Step 2: Collect equipment characteristics for the facilities.

Step 3: Generate engineering based load shapes for non-thermal loads in the facilities.

Step 4: Generate load shapes for the thermal [heating, ventilation, and air conditioning (HVAC)] loads in the facilities.

Step 5: Aggregate end-use data to the desired level.

Thermal loads consist of loads that vary with the solar exposure of the building, time of day, ambient weather conditions, and the internal loads in the building. End-uses that are commonly considered to be thermal loads include the HVAC systems and the refrigeration systems used for food storage. The non-thermal loads in facilities consist of loads that run without regard for the external weather conditions, but rather are run according to a fixed operating schedule. Examples of non-thermal load include indoor lighting, personal computer equipment, and parking lot lights. Although the non-thermal loads are relatively easy to estimate, more complex methods are required to estimate the thermal loads. There are several building simulation models, such as DOE-2, that can be used to estimate the load shapes for thermal end uses in a facility.

Statistical Models

Although there are several types of statistical models that can be used to develop end-use load shapes, one of the most straightforward approaches is called the conditional demand analysis (CDA) method. Another common approach that combines statistical analysis with engineering models is called the statistically adjusted engineering model and will be introduced later in this chapter. The CDA model is based on the premise that the sum of the electrical demand consumed by the end uses in the facility at any time is equivalent to the electrical demand consumed by the entire facility, as shown in Equation 2-7. Fortunately, most utilities already collect whole-building load research data for a sample of customers in their service territory. If whole-building data is not currently available, whole-building meters can be installed and monitored at a much lower cost than installing end-use meters in buildings.

$$L_{ht} = \sum_{i=1}^{n} L_{hi}$$ (Equation 2-7)

where:

L_{ht} is the hourly whole-building load research data for the facility at time h

L_{hi} is load consumed by end use i at time h.

Since the end-use loads are not known for the facility, estimates can be used in place of the actual values. In their simplest form, these estimates can simply be indicators of whether or not the end use is present at the facility and consuming electricity for the time period in question, where a zero represents that an end use is not present or not consuming electricity and a one indicates that the end use is present and consuming electricity. The CDA equation for this case is shown in Equation 2-8.

$$L_{ht} = Ind_{h1} + Ind_{h2} + \ldots + Ind_{hn}$$ (Equation 2-8)

where:

L_{ht} is the hourly whole-building load research data for the facility at time h

L_{hi} is the indicator for end use i.

The CDA model can be made more accurate by replacing the indicator value with the product of the floor space used by the end use and an estimate of the energy use index (EUI) for the facility, as shown in Equation 2-9.

$$E_{ht} = \sum_{i=1}^{n} (FS_{hi})(EUI_{hi})$$ **(Equation 2-9)**

where:

E_{ht} is the energy consumed by the facility over hour h

FS_{hi} is the floor space occupied by end-use i

EUI_{hi} is the energy-use index for end-use i at hour h

EUIs represent the amount of energy consumed per floor space and are generally given in units of kilowatt-hours per square foot (kWh/ft^2) or thousands of British Thermal Units (BTU) per square foot ($MBTU/ft^2$). The steps required for statistical analysis are:

Step 1: Sample design.

Step 2: Collect end-use characteristics for the facilities.

Step 3: Develop statistical models.

Step 4: Apply statistical results to produce end-use loads.

Step 5: Aggregate end-use data to the desired level.

The actual form of the equation used in the regression analysis is shown in Equation 2-10.

$$E_{ht} = \beta_{h1} \cdot E_{h1} + \beta_{h2} \cdot E_{h2} + \ldots + \beta_{hn} \cdot E_{hn}$$ **(Equation 2-10)**

where:

E_{ht} is the energy consumed by the facility over hour h

E_{h1} is the indicator or estimate for end use 1

E_{h2} is the indicator or estimate for end use 2, through end use n.

These values are provided as input to the statistical model. The model results are in the form of statistical coefficients (βs) where β_1 is the coefficient for end-use estimate 1, etc.

To apply the results of the statistical model the resulting β's are multiplied by the corresponding estimate or indicator for each end use to develop a final result. One typical method of generating the statistical model is to run the statistical analysis for each facility and hour of the day across all of the days of the year.

Statistical/Engineering Models

As the name of this category implies, these models combine elements of both the engineering models and statistical models. In fact, it can sometimes be difficult to differentiate between these types of models. One of the more common statistical/engineering approaches is called the statistically adjusted engineering (SAE) model. With the SAE model, Equation 2-10 can be utilized where the end-use estimates are developed from an engineering model prior to running the statistical model. The resulting coefficients (β's) are then inverted and multiplied by the engineering estimates to provide final end-use estimates. The following steps are required for the SAE model:

Step 1: Sample design.

Step 2: Collect data for the facilities.

Step 3: Develop engineering estimates.

Step 4: Develop statistical models.

Step 5: Apply statistical results to produce end-use loads.

Step 6: Aggregate end-use data to the desired level.

Reference

"Glossary of Terms for the Electric Power Industry," Electric Power Training Center, Western Area Power Administration, 1990.

3
Sample Design

What Is Sample Design?

The purpose of sample design is to select a group of customers that have some parameter that is representative of a larger group of customers. However, before the sample design is begun several questions concerning the objectives of the study need to be addressed including:

- What is your desired target population?
- What is the sample frame?
- Which characteristics (target variables) of the population are you most interested in?
- What are your accuracy expectations and resource constraints?

Target Population

To define the target population, you need to decide who you want included in your population as well as the unit of analysis for the study. In a residential study some potential target populations include single-family homes, duplex homes, or multi-family homes. You may decide to include all, some, or only one of these categories in your target population. If the desired unit of analysis is defined as a dwelling where one family resides, the single-family home would constitute a single unit of analysis, a duplex would contain two dwellings, and a multi-family home would contain several dwellings. Contrast this with defining the unit of analysis as all space enclosed by a building shell. In the latter case the single-family home, duplex, and multi-family home would all be considered as a single unit of analysis. The desired results of the analysis will partially dictate the unit of analysis. For instance, if you want an estimate of the total number of domestic refrigerators in your service territory, you could use either of the above definitions for the unit of analysis. However, if you want an estimate of the average number of refrigerators and the corresponding average energy consumption among the families in your service territory, you probably need to define the unit of analysis as a single-family dwelling. For a study of commercial buildings some of the available choices for unit of analysis are the economic unit where business is conducted, a building or group of buildings controlled by a single decision maker, and an electric or gas meter.

Sample Frame

Defining the target population is the first step in developing the sample frame. Other issues that need to be addressed when developing the sample frame are (Dohrmann, D. R., and Alereza, T., September 1984):

- Are the units of analysis in the target population frame defined the same in the sample frame?
- Are the units in the sample frame not in the target population?
- Are the units in the target population not in the sample frame?

One common problem that occurs with the sample design is that the units may have a different definition in the target population than in the sample frame. For example, if you defined the unit to be single-family dwellings in the target population does that agree with the unit definition in the sample frame? If the sample frame is defined by electric and/or gas meters there are generally two ways that the meters are installed. In the first case, a single meter is used to measure the energy use at the building level. In this case, the single unit in the sample frame contains multiple units in the target population. If, however, each dwelling unit were individually metered then the unit definition in the sample frame and target population would be consistent. When you look at commercial and industrial customers you can get even more complex. In addition to having several units connected to a single meter, you may encounter cases where a single unit in the target population is served by several meters (or multiple units) in the sample frame. Some of the most complex cases occur when you have several units in the target population corresponding to several units in the sample frame, which is frequently the case with college campuses and shopping malls.

Units that are not designed to be in the target population but are in the sample frame are considered out-of-scope units and should be eliminated if it is cost effective. If your target population was limited to sampling commercial and industrial establishments, out-of-scope accounts might include electrical substations, street lights, and private residences.

If the desired units in the target population are missing from the sample frame you have two choices, either eliminate those missing units from the target population or supplement the sample frame with data from other sources.

Target Variable

The selection of the target variable also affects the choice of which sample design scheme is used to develop the sample. Some questions that need to be addressed for the target variable include (EPRI, 1984):

- What will you be estimating from the data?
- Do you have multiple estimate goals?
- Do you want a single estimate for the entire population or estimates for individual groups in the population?

For the desired estimates, it is best if you perform the sample design to capture significant differences in the estimates. For example, if you wanted to calculate the energy use intensity across groups in the population you may want to sample by building classification. However, if you are interested in examining how computers are utilized in the population you might want to sample uniformly across your population according to how many computers are used at each site, if that is known.

However, in any large analysis project there will probably be multiple goals, each of which would have its own preferred sampling scheme. Therefore, it is probably desirable to define the single most important parameter to be estimated and base the sample design on that selection. Any secondary estimates will then be limited to whatever data is available in the primary sample.

To actually implement the sample design you need to choose a target variable to base the design on. Since electric utilities have traditionally been concerned with energy use or electrical demand, many of their studies have used either monthly billing history or monthly peak demand as a target variable. The important thing to keep in mind when choosing the target variable is that it should be correlated with whatever parameters you will be estimating. Some items that tend to be strongly correlated for energy use in building are monthly peak demand and monthly energy use and annual energy use and floor space.

If possible, it is recommended that you check to ensure your target variables and the primary parameters you intend to estimate are correlated before you perform the sample design. One source for this test data may be studies that have previously been performed on your target population.

Statistical Relationships

Another important aspect of the sample design is that it allows the utility to study the interactions between sample size, precision, and confidence level. If the utility wants to increase the statistical confidence of the load shapes, it can increase the number of units being analyzed. Conversely, if it doesn't care as much about the precision of the final load shapes, it can chose to have fewer units in the sample. Equation 3-1 presents these interactions and can be used to determine how many points are required for a simple sample from a population.

$$n = \frac{N^2 \cdot S_x^2}{\sigma_x^2 + N \cdot S_x^2} \qquad \text{(Equation 3-1)}$$

where:

n = sample size

N = population size

σ_x^2 = squared standard error of the design variable in the population and

S_x^2 = standard dispersion of the design variable in the population.

Standard Error

The Standard Error is calculated as the square of the product of the expected precision and the sum of the design variable divided by the confidence level, as presented in Equation 3-2:

$$\sigma_x = \frac{\varepsilon \cdot X}{Z} \qquad \text{(Equation 3-2)}$$

where:

ε = desired precision for the sample

X = sum of the design variable in the population and

Z = desired confidence level for the sample.

Confidence Level

The confidence level (Z) is defined as the probability that the estimate will fall within the one-tailed normal distribution of estimates a certain percentage of the time. Table 3-1 presents some commonly used Z values along with their corresponding probabilities.

Table 3-1. Distribution probabilities and their corresponding Z values.

Probability	Z Value
95%	1.96
90%	1.64
85%	1.44
80%	1.28
75%	1.15
70%	1.04

Coefficient of Variation

The coefficient of variation (CV) of variable X is calculated by dividing the standard deviation of the estimate by the mean of the estimate as shown in Equation 3-3.

$$CV_x = \frac{S_x}{\overline{X}}$$ (Equation 3-3)

where:

S_x = standard deviation of variable X and

$\overline{X}$ = mean of variable X.

Mean

The arithmetic mean of variable X is calculated by dividing the sum of the values by the number estimates as shown in Equation 3-4.

$$\overline{X} = \frac{\sum_{i=1}^{n} X_i}{n}$$ (Equation 3-4)

where:

n = number of values to be averaged and

X_i = value of variable X for observation i.

Standard Deviation

The standard deviation (S) of variable X is calculated by taking the sum of the squared values minus the product of the number of observations and the square of the mean, dividing the sum by the number of observations, and taking the square root as shown in Equation 3-5.

$$S_x = \sqrt{\frac{\sum_{i=1}^{n}(X_i)^2 - n \cdot \overline{X}^2}{n}}$$ (Equation 3-5)

where:

X_i = value of variable X for observation i and
$\overline{X}$ = mean value of variable X.

Standard Dispersion

The standard dispersion (S_x^2) is calculated as the square of the standard deviation of variable X.

Sampling Techniques

There are several sample design schemes that have traditionally been utilized to develop samples for analyzing the energy use of a population. Three of the most common sampling techniques used are simple random sampling, sampling with uniform stratification, and sampling with optimal stratification.

Simple Random Sampling

Simple random sampling is useful when you want to collect characteristics of your population as a whole and are not concerned that categories within the sample have a predefined statistical accuracy. Equation 3-1 can be utilized to determine the required sample size for a given population.

Example 3-1: Simple Random Sample
For this example, we will perform the sample design utilizing the average monthly energy use as the design variable for a generic

population of buildings, shown in Table 3-2. From the data in Table 3-2, we are able to calculate the following statistics for the population:

$\Sigma X = 2{,}352{,}332$

$N = 80$ and

$S_x^2 = 4.705 \times 10^8$

Assume that we want a precision of 15 percent and a Z value corresponding to a 90 percent probability. From Table 3-1, the selected Z value is 1.64.

Substituting these values into Equation 3-2, we obtain:

$$\sigma_x^2 = \left(\frac{0.15 \cdot 2{,}352{,}332}{1.64}\right)^2$$

$$= 4.629 \times 10^{10}$$

Next, we substitute the required values into Equation 3-1 to obtain:

$$n = \frac{80^2 \cdot 4.705 \times 10^8}{4.629 \times 10^{10} + 80 \cdot 4.705 \times 10^8}$$

$$= 36 \text{ accounts}$$

To meet these conditions, we would need to sample 36 out of the 80 buildings, or slightly more than half of the population. What happens if we increase the precision requirements to 10 percent? We would need to sample 52 of the buildings.

Example 3-2: Expanded Random Sample

Now let's study a more realistic example of a population of buildings from a generic utility. In this case, we will examine developing a single sample for entire commercial and industrial population for the generic utility. The following statistics were generated for the overall population based on the average energy used by account per month:

$N = 3{,}050$ accounts

$\overline{X} = 10{,}177$ kWh/month

Table 3-2. Generic population.

ID	kWh/month	ID	kWh/month
1	523	41	29,083
2	621	42	30,504
3	821	43	30,894
4	903	44	31,367
5	1,191	45	31,609
6	1,637	46	33,636
7	1,950	47	33,772
8	3,087	48	33,811
9	3,113	49	34,268
10	4,423	50	34,719
11	4,752	51	35,198
12	4,788	52	37,009
13	6,147	53	37,665
14	6,172	54	39,248
15	6,951	55	39,756
16	7,246	56	40,180
17	7,785	57	41,037
18	7,900	58	42,883
19	9,516	59	43,060
20	9,623	60	43,314
21	11,688	61	43,911
22	12,302	62	45,354
23	12,400	63	46,503
24	13,163	64	48,413
25	13,368	65	51,352
26	14,099	66	52,099
27	14,424	67	53,238
28	14,750	68	54,340
29	14,803	69	55,195
30	15,011	70	55,434
31	15,791	71	56,514
32	16,693	72	58,564
33	17,947	73	60,170
34	18,859	74	65,340
35	20,058	75	70,863
36	23,337	76	72,765
37	23,414	77	73,479
38	24,158	78	74,798
39	25,600	79	75,801
40	28,641	80	75,942

$S_x^2 = 3.941 \times 10^8$ (kWh/month)^2 and
$X = N \cdot \overline{X} = 3.104 \times 10^7$ kWh/month.

Next, assume that we want to find the sample size for this population with a 10 percent precision and a 95 percent confidence. Using Equation 3-2, we obtain:

$$\sigma_x^2 = \left(\frac{0.10 \cdot 3.104 \times 10^7}{1.96}\right)^2$$

$$= 2.508 \times 10^{12}$$

Next, substitute these results into Equation 3-1 to find the required sample size for this population:

$$n = \frac{3,050^2 \cdot 3.941 \times 10^8}{2.508 \times 10^{12} + 3,050 \cdot 3.941 \times 10^8}$$

$$= 988 \text{ accounts}$$

Stratified Sampling

Upon examination of Equation 3-1, it becomes clear that in order to reduce the required sample size you either need to reduce the desired standard error of the sample or reduce the standard dispersion in the sample. Stratifying a population before the sample design stage is one method that can be used to reduce the dispersion within a sample. To stratify a sample you divide the sample into multiple groups which are fairly homogenous with respect to the design variable and draw a random sample from each group. This reduces the standard deviation and standard dispersion within each group. In this section we will look at some of the methods that can be used for stratifying utility accounts before performing the sample design including uniform stratification, stratification by building type, and optimal stratification. The final scheme that we will examine is using proportional allocation to stratify a population after the sample design has been performed.

Uniform Stratification

Uniform stratification is implemented by dividing the population into groups (or strata) of equal intervals. Equation 3-1 is then applied to each strata to determine the number of sample points required within each strata. The total number of sample points required is obtained by summing the number of points needed within each strata.

Example 3-3: Uniform Stratification

Calculate the number of points required in each strata for the utility population presented in Example 3-2. For this analysis, we chose to split the population into four equal strata and generated the following data:

Strata (h)	Lower kWh	Upper kWh	N_h	$\overline{X}_h$ (kWh/month)	S_x^2 (kWh/month)	X_h (kWh)
1	0	<40,500	2,793	5,087	6.938×10^7	1.421×10^7
2	40,500	<81,000	211	56,515	1.435×10^8	1.192×10^7
3	81,000	<121,500	34	95,542	1.776×10^8	3.248×10^6
4	121,500	<162,000	12	138,270	1.591×10^8	1.659×10^6

Using these data, we calculated the required standard error using Equation 3-2 with a confidence level of 95 percent and a precision of 10 percent. For strata one, X is 1.421×10^7 kWh per month and Z is 1.96. Substituting these values into Equation 3-2 gives:

$$\sigma_x^2 = \left(\frac{0.10 \bullet 1.421 \times 10^7}{1.96} \right)^2$$

$$= 5.256 \times 10^{11}$$

The results for squared standard error are listed below for each of the strata:

Strata	σ_x^2
1	5.256×10^{11}
2	3.699×10^{11}
3	2.746×10^{10}
4	7.164×10^9

As a final step, these results are used in conjunction with Equation 3-1 to obtain the sample size for each strata. For strata 1, N is 2,793, and S_x^2 is 6.938×10^7. Substituting these values into Equation 3-1 yields:

$$n = \frac{2,793^2 \cdot 6.938 \times 10^7}{5.256 \times 10^{11} + 2,793 \cdot 6.938 \times 10^7}$$

$$= 752 \text{ accounts}$$

The sample sizes for all of the strata are listed below:

Strata	n_h
1	752
2	16
3	6
4	3
Overall	777

For the uniform stratification, we require only 777 units compared to 988 for random sampling with no stratification, which is approximately a 20 percent reduction in the number of sample points required.

Stratification by Building Type

If you are only interested in estimating parameters for the entire population then either the random selection or uniform stratification over the entire population works fine. You can even estimate parameters for different groups of customers, but the estimates may not be representative of customer groups within your population or have a high statistical confidence. One method that can be used to help ensure a reasonable representation of those customer groups and a predetermined statistical confidence is to stratify the population into the desired customer groups. It is common for utilities to either stratify their customers by building type or rate class. To illustrate how to stratify by building type, we will look at another example.

Example 3-4: Stratification by Building Type

Calculate the number of points required in each building type if the utility population presented in Example 3-1 is stratified by building type. For this analysis, we calculated the following data:

Building Type (h)	N_h	$\overline{X}_h$ (kWh/month)	S_x^2 (kWh/month)^2	X_h (kWh)
Office	900	10,338	3.584×10^8	9.304×10^6
Restaurant	300	12,090	2.944×10^8	3.627×10^6
Retail	700	13,327	5.331×10^8	9.329×10^6
Grocery	180	16,917	9.456×10^8	3.045×10^6
School	30	34,416	2.378×10^9	1.032×10^6
Health	60	14,501	5.787×10^8	8.700×10^5
Warehouse	320	2,240	7.198×10^6	7.168×10^5
Miscellaneous	540	4,804	9.074×10^7	2.594×10^6
Industrial	20	26,089	7.320×10^8	5.218×10^5

Using these data, we calculated the required standard error using Equation 3-2 with a confidence level of 95 percent and a precision of 10 percent. For the Office strata, X is 9.304×10^6 kWh per month and Z is 1.96. Substituting these values into Equation 3-2 gives:

$$\sigma_x^2 = \left(\frac{0.10 \bullet 9.304 \times 10^6}{1.96} \right)^2$$

$$= 2.253 \times 10^{11}$$

The results for the squared standard errors are listed below for all of the strata:

Strata	σ_x^2
Office	2.253×10^{11}
Restaurant	3.424×10^{10}
Retail	2.265×10^{11}
Grocery	2.414×10^{10}
School	2.772×10^9
Health	1.970×10^9
Warehouse	1.337×10^9
Miscellaneous	1.752×10^{10}
Industrial	7.088×10^8

Substituting these values into Equation 3-1 to obtain the sample size for the office strata yields 530 units required as shown below:

$$n = \frac{900^2 \cdot 3.584 \times 10^8}{2.253 \times 10^{11} + 900 \cdot 3.584 \times 10^8}$$

$$= 530 \text{ accounts}$$

The required sample sizes for all of the strata are:

Strata	n_h
Office	530
Restaurant	216
Retail	436
Grocery	158
School	29
Health	57
Warehouse	202
Miscellaneous	398
Industrial	19
Overall	2,044

Therefore, it would require sampling 2,044 units out of a population of 3,050 units, or 67 percent of the population, to obtain the desired accuracy using random sampling within each strata. Obviously this would probably place large resource constraints on any utility.

As we have already seen, the two most direct methods of reducing the required sample sizes are to either relax the accuracy requirements of any estimates or perform some sort of stratification scheme upon the population. If we were to relax the precision requirements to 15 percent and require a z value for 85 percent accuracy for the building types presented in Example 3-4, we would obtain the following sample sizes:

Strata	n_h
Office	277
Restaurant	134
Retail	237
Grocery	124

Strata	n_h
School	27
Health	51
Warehouse	112
Miscellaneous	251
Industrial	17
Overall	1,229

The total number of sample units was cut almost in half by changing the statistical accuracy requirements. In the next two sections, we will examine how stratifying the population within each building type affects our required sample size.

Uniform Stratification Within Building Type

For uniform stratification within the building types we treat each of the building type categories as a unique domain and perform a uniform stratification scheme for each of the domains.

Example 3-5: Uniform Stratification by Building Type

If we perform a uniform stratification scheme on the building types presented in Example 3-4, using four strata per building type, a 10 percent confidence level, and a Z value representing a 95 percent probability, we obtain the results presented in Table 3-3. It is important to realize that some of the strata are calculated to have a sample size of less than one or zero. However, we have made a decision that each of the strata we have selected will have a minimum of one sample unit.

From the results presented in Table 3-3, it appears that the total number of sampling units required is 1,651 compared to 2,044 when we simply stratified the population by building type. It is important to realize that we are comparing apples to oranges for these two cases since in the case of the simple random stratification we assumed a precision of 10 percent and confidence level of 95 percent for the population of office buildings, but in the case of the uniform stratification we assumed the same precision and confidence levels for each individual strata in the office buildings which results in a lower standard error for the office buildings.

Table 3-3. Results for uniform stratification within building types.

Building Type	h	N_h	$\overline{X}_h$	S_x	S_x^2	X	σ_x^2	n_h
Office	1	741	2,489	4,010	1.608×10^7	1.844×10^6	8.943×10^9	423
	2	69	30,152	5,745	3.301×10^7	2.080×10^6	1.138×10^{10}	12
	3	50	49,299	5,780	3.340×10^7	2.465×10^6	1.598×10^{10}	5
	4	40	72,876	6,402	4.098×10^7	2.915×10^6	2.235×10^{10}	3
Restaurant	1	238	4,441	4,158	1.729×10^7	1.057×10^6	2.938×10^9	139
	2	33	29,443	5,704	3.254×10^7	9.716×10^5	2.483×10^9	10
	3	22	50,571	5,928	3.514×10^7	1.113×10^6	3.255×10^9	4
	4	6	77,540	5,043	2.543×10^7	4.652×10^5	5.692×10^8	1
Retail	1	588	4,656	7,253	5.261×10^7	2.738×10^6	1.971×10^{10}	359
	2	75	44,201	7,330	5.373×10^7	3.315×10^6	2.890×10^{10}	9
	3	24	76,382	8,753	7.661×10^7	1.833×10^6	8.838×10^9	4
	4	13	110,963	10,056	1.011×10^8	1.443×10^6	5.472×10^9	3
Grocery	1	158	6,871	9,116	8.311×10^7	1.086×10^6	3.099×10^9	128
	2	11	60,765	10,834	1.174×10^8	6.684×10^5	1.175×10^9	6
	3	6	96,717	14,395	2.072×10^8	5.803×10^5	8.856×10^8	4
	4	5	142,137	12,909	1.666×10^8	7.107×10^5	1.328×10^9	2

Category							
School	1	23	8,729	5,865	3.440×10^7	2.008×10^5	1.060×10^8	20
	2	1	69,358	0	0.000×10^0	6.936×10^4	1.265×10^7	1
	3	2	103,733	10,710	1.147×10^8	2.075×10^5	1.132×10^8	1
	4	4	138,725	14,169	2.008×10^8	5.549×10^5	8.098×10^8	2
Health	1	54	7,975	9,742	9.491×10^7	4.306×10^5	4.877×10^8	49
	2	3	47,563	8,641	7.467×10^7	1.427×10^5	5.354×10^7	2
	3	2	78,818	856	7.329×10^5	1.576×10^5	6.535×10^7	1
	4	1	139,072	0	0.000×10^0	1.391×10^5	5.086×10^7	1
Warehouse	1	231	835	728	5.294×10^5	1.930×10^5	9.792×10^7	128
	2	56	4,284	920	8.459×10^5	2.399×10^5	1.513×10^8	13
	3	23	7,566	861	7.416×10^5	1.740×10^5	7.963×10^7	4
	4	10	10,998	865	7.474×10^5	1.100×10^5	3.181×10^7	2
Miscellaneous	1	482	2,009	2,748	7.551×10^6	9.684×10^5	2.466×10^9	287
	2	39	19,751	3,019	9.113×10^6	7.703×10^5	1.560×10^9	7
	3	10	36,605	4,792	2.296×10^7	3.660×10^5	3.524×10^8	4
	4	9	54,357	3,056	9.339×10^6	4.892×10^5	6.294×10^8	1
Industrial	1	11	5,904	5,182	2.686×10^7	6.494×10^4	1.109×10^7	11
	2	3	25,830	3,969	1.576×10^7	7.749×10^4	1.579×10^7	2
	3	3	48,051	1,944	3.778×10^6	1.442×10^5	5.465×10^7	1
	4	3	78,401	6,990	4.886×10^7	2.352×10^5	1.455×10^8	2

Optimal Stratification Within Building Type

While the uniform stratification scheme is an improvement over a random sample within each building type, there are still more schemes that will help optimize the strata break points and the number of points required per strata. One commonly used set of optimization tools are the Delenius-Hodges stratification scheme with Neyman allocation.

Delenius-Hodges Stratification

In the uniform stratification scheme, strata break points are defined so they occur in equal blocks. In the Delenius-Hodges stratification scheme the strata break points are defined using the following steps:

1. Define equal sized bins for the population.

2. Determine the frequency of population units in each bin.

3. Take the square root of the frequency in each bin.

4. Determine the cumulative square root of the frequency in each bin.

5. Determine the number of strata you want in the domain.

6. Define the strata break points at equal cumulative square root values.

Example 3-6: Delenius-Hodges Stratification

For this example, we will treat the office buildings in the generic utility population as our domain to stratify. The first step is to define the size of the bins. We chose bins of 2,500 kWh of average monthly energy use each. Therefore, bin 1 contains accounts which had an average energy use from 0 to 2,500 kWh, bin 2 contains accounts with an energy use of 2,500 to 5,000 kWh per month, and so on. This worked out to give us approximately 33 bins for the office buildings. Remember, the smaller the bins you define the more calculations you will have to perform. The results of steps 2, 3, and 4 are shown in Table 3-4.

Table 3-4. Stratification of office buildings using the Delenius-Hodges method.

Bin	F	$\sqrt{F}$	Cumulative $\sqrt{F}$	Strata
2,500	557	23.601	23.601	1
5,000	67	8.185	31.786	2
7,500	38	6.164	37.951	2
10,000	20	4.472	42.423	2
12,500	20	4.472	46.895	2
15,000	16	4.000	50.895	2
17,500	15	3.873	54.768	2
20,000	8	2.828	57.596	2
22,500	9	3.000	60.596	3
25,000	6	2.449	63.046	3
27,500	12	3.464	66.510	3
30,000	3	1.732	68.242	3
32,500	14	3.742	71.984	3
35,000	10	3.162	75.146	3
37,500	6	2.449	77.595	3
40,000	7	2.646	80.241	3
42,500	8	2.828	83.070	3
45,000	8	2.828	85.898	3
47,500	7	2.646	88.544	3
50,000	12	3.464	92.008	4
52,500	1	1.000	93.008	4
55,000	3	1.732	94.740	4
57,500	7	2.646	97.386	4
60,000	5	2.236	99.622	4
62,500	3	1.732	101.354	4
65,000	5	2.236	103.590	4
67,500	4	2.000	105.590	4
70,000	3	1.732	107.322	4
72,500	2	1.414	108.736	4
75,000	6	2.449	111.186	4
77,500	6	2.449	113.635	4
80,000	6	2.449	116.085	4
82,500	6	2.449	118.534	4

Next, we assumed that we wanted four strata in the office building domain. Keep in mind that this is actually an iterative process where you would try different numbers of strata to determine which one produces the optimal number of sample units. The total cumulative value of the square root of the frequencies was 118.534. Therefore, the break point between strata 1 and strata 2 was calculated by dividing 118.534 by 4, equaling 29.634. The break points for each of the strata are listed below:

Strata	Cumulative $\sqrt{F}$ Break Point
1	29.634
2	59.267
3	88.901
4	118.534

Neyman Allocation Scheme

The Neyman allocation scheme distributes the sample points among the strata based on the number of points in the population and the standard deviation in each strata. Equation 3-6 shows how to calculate the number of sample points required in each strata using the Neyman allocation scheme.

$$n_h = n \cdot \frac{N_h S_h}{\sum_{h=1}^{s} N_h S_h} \qquad \textbf{(Equation 3-6)}$$

where:

n = number of sample points in the domain

n_h = number of sample points in strata h

N_h = number of units in the domain in strata h

S_h = standard deviation of the domain in strata h and

s = number of strata.

The total number of sample points required by all of the strata can be found using Equation 3-7 as shown on the following page:

$$n = \frac{\left(\sum_{h=1}^{s} N_h S_h\right)^2}{\sigma_x^2 + \sum_{h=1}^{s} N_h S_h^2}$$ (Equation 3-7)

where:

n = number of required sample points in the overall population

N = number of units in the population

N_h = number of units in population strata h

S_h = standard deviation in strata h

S_h^2 = standard dispersion in strata h and

σ_x^2 = squared standard error of the mean design value.

In Example 3-5, we examined the Delenius-Hodges stratification scheme to determine the stratum break points. Now let's examine how to use the Neyman allocation to distribute the sample points among the strata.

Example 3-7: Neyman Allocation

Using the data for the office buildings from Example 3-5, we calculated the following statistics:

Strata	N_h	S_h	S_h^2
1	557	499	2.495×10^5
2	184	4,730	2.237×10^7
3	90	7,760	6.022×10^7
4	69	11,221	1.259×10^8

In addition, we calculate the squared standard error using Equation 3-1 with a population total monthly energy use of 9.304×10^6 kWh per month, a precision of 10 percent, and a Z probability of 95 percent as follows:

$$\sigma_x^2 = \left(\frac{0.1 \cdot 9.304 \times 10^6}{1.96}\right)^2$$

$$= 2.253 \times 10^{11} \text{ (kWh per month)}^2$$

Next, we utilized Equation 3-7 to find the required sample size for all of the strata using the calculation results presented below:

$$\left(\sum_{h=1}^{s} N_h S_h\right)^2 = 6.869 \times 10^{12} \text{ (kWh per month)}^2$$

$$\sum_{h=1}^{s} N_h S_h^2 = 1,836 \times 10^{10} \text{ (kWh per month)}^2$$

$$n = \frac{6.869 \times 10^{12}}{2.253 \times 10^{11} + 1.836 \times 10^{10}}$$

$$= 28$$

Recall that in Example 3-4 we determined we needed a sample size of 530 units for the office building sector when utilizing a simple random sample scheme.

As a final step, we applied Neyman allocation, using Equation 3-6, to distribute the sample points among the strata. The results of the Neyman allocation are listed below:

Strata	$N_h S_h$	n_h
1	2.779×10^5	3
2	8.703×10^5	9
3	6.984×10^5	8
4	7.742×10^5	8
Overall	2.621×10^6	28

Proportional Allocation

An alternative to using the Neyman allocation is to use proportional allocation where each strata is allocated a portion of the total required sample points for the population according to Equation 3-8 as shown below:

$$n_h = n \cdot \frac{N_h}{N} \qquad \textbf{(Equation 3-8)}$$

Before using Equation 3-8, you will need to determine the overall sample size for the population (n) using either simple random sampling or a stratification scheme of your choosing.

Reference

Dohrmann, D.R., and Alereza, T., "Sampling Methodologies for the Commercial Sector," EPRI EA-3688, Electric Power Research Institute, Final Report, September 1984.

4
End-Use Metering

What Is End-Use Metering?

For this book, end-use metering refers to the act of installing any type of metering on an appliance, electrical circuit, or building meter to determine how much energy is being consumed. The fuel consumption of the device(s) can be measured either instantaneously or over a long period of time. When electricity is measured instantaneously, it is referred to as electrical demand and the most common units are kilowatts (kW). However, when electricity is measured over a period of time, it is referred to as electricity usage and the units of measurement are kilowatt-hours (kWh). It is common for residential electricity users to be charged only for kWh usage, while industrial and commercial customers are commonly charged for both kWh and kW. Natural gas is most commonly measured in terms of how many cubic feet of natural gas flow through the meter in a given period of time.

Electrical Demand and Electricity Usage

When metering electrical equipment or using metered data in analysis, it is important to understand the difference between instantaneous demand, average demand, and energy use. Instantaneous demand is the maximum demand that occurs during a period of time while average demand is the mean value of a group of instantaneous demand points over a period of time. Energy use is the cumulative total of instantaneous demand points

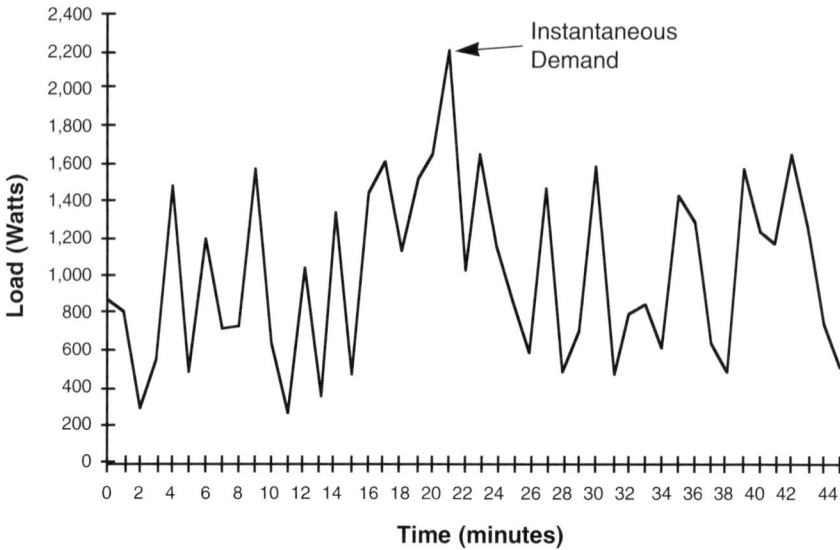

Figure 4-1. Demand variation over time.

over a period of time. Figure 4-1 presents a graphical depiction of instantaneous demand while Figure 4-2 depicts average demand over a period of time.

A graphical depiction of cumulative energy use is presented in Figure 4-3.

Energy use can be mathematically represented as the sum of the instantaneous demand values over a period of time as shown in Equation 4-1.

$$energy = \sum_{i=0}^{t} D_i \qquad \textbf{(Equation 4-1)}$$

where:

energy = energy use over time t

D_i = instantaneous demand at time i

A lot of power meters that have recording capability also have the ability to select what period of time the demand data is averaged over or even the ability to report a moving average demand. Figure 4-3 graphically depicts a moving average along with the demand data previously presented in Figure 4-2. The moving average

End-Use Metering 55

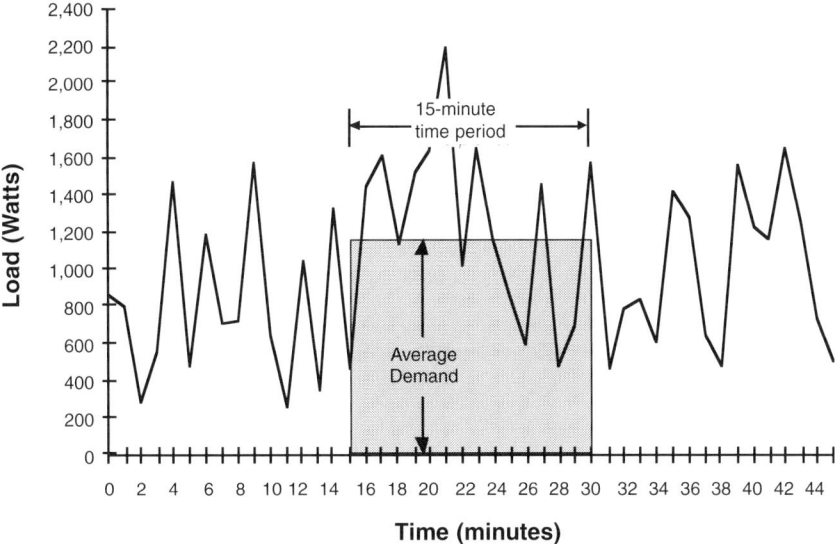

Figure 4-2. Time period to average.

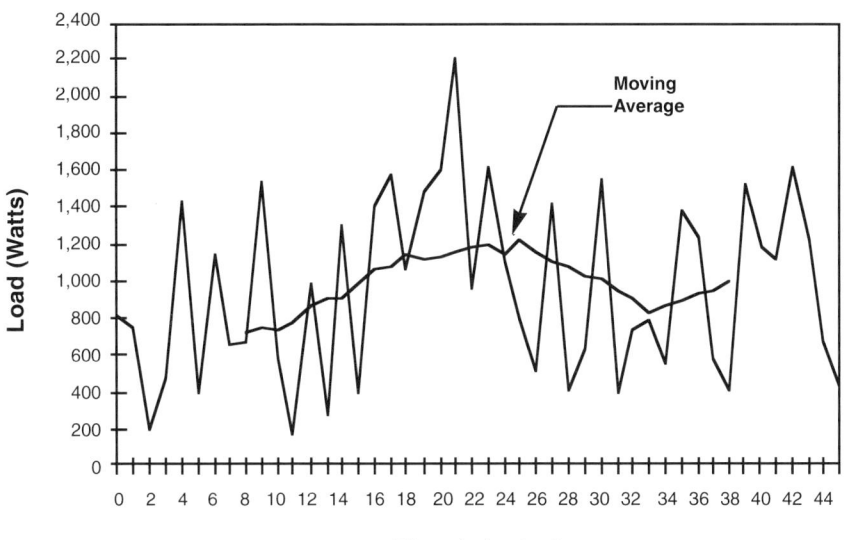

Figure 4-3. Moving average demand.

presented in Figure 4-3 was calculated by averaging the demand for a period of 15 minutes using a given point in time as the midpoint for the averaging process. For example, to find the moving average demand at minute 8, the electrical demands from minute 1 to minute 15 were averaged together; to calculate the moving demand at minute 9, the electrical demand from minute 2 to minute 16 were averaged together, etc.

The most important thing to remember with metered data is that if the data is averaged as opposed to using instantaneous data, you are left with less information. Whether or not this is a concern to you depends on what you are using the data for.

For example, if you are examining the operation of a set of compressors in a grocery store, you probably would want to use instantaneous demand values so you could examine the cycling patterns of the compressors. However, if you are studying the total load in the grocery store, the instantaneous data would probably have more "noise" or variation than you need for your purposes, and you might be satisfied using average demand or moving average demand data.

Types of Meters

You have a project that you want to collect metered data on, but what kind of meter should you use? There are several types of meters currently on the market including:

- voltage/ampere meters
- power meters
- energy meters
- pulse meters
- flow meters
- run-time meters

Voltage and Ampere Meters

Voltage and ampere meters are some of the most inexpensive meters on the market. Hand-held multi-meters and data loggers commonly measure voltage.

Power Meters

For electrical equipment, power meters have traditionally been the most accurate and the most expensive type of meter for collecting end-use information. For a meter to record the power flowing into the meter, the meter needs to measure both the voltage and current in each phase and calculate the equivalent power rating. Power measurements can now be made with virtually all ranges of metering equipment from hand-held multi-meters through multi-channel meters.

Energy Meters

Energy meters are capable of recording either cumulative energy use over the metering period or energy use by period of the day. The most common form of energy meters are electric and/or gas meters which can be seen on most buildings.

Run-Time Meters

Run-time meters are used to record the hours of operation that a device is on, the cumulative hours that a device is on, or even the cumulative hours that a device is on by period of day or week. There are several dedicated run-time meters available on the market.

Flow Meters

Flow meters are used to measure fluids (such as natural gas) that flow through a pipe. Flow meters can either be internal devices, such as in-line turbines, or external meters, such as sonic flow meters.

Metering Costs

The cost of meters can range from a hundred dollars to several thousand dollars. Table 4-1 presents a matrix of types of meters, capabilities, and cost ranges (Bowman and Goldberg, 1994).

Table 4-1. Typical metering capabilities and costs.

Meter Type	Power	Amps	Voltage	Hours	Cost Range ($)
Hand-held Multi-meters	•	•	•		150–1200
Three-phase power analyzers	•	•	•		3,000–15,000
Runtime loggers 1 to 4 channel				•	100–200
Data loggers multi-channel	•	•	•	•	800–2,000
Data loggers Multi-channel	•	•	•	•	800–4,000
Meter/recorders		•			1,000–5,000

Choosing the Correct Meter for the Job

If you have an unlimited budget, you will probably want to buy the best power meters available on the market. However, if you're like most people you have limited budget with which to produce the most accurate results possible. This section will provide guidelines for selecting the most cost-effective metering equipment to collect end-use load shape data for constant load, variable load, and HVAC applications.

Constant Load With Fixed Hours

A constant load with fixed hours is where the load is turned on and off according to a predetermined schedule. Examples of such a load might be parking lot lights or street lights with photosensors. In this case, the most cost-effective approach is to measure the power being drawn by the loads when they are on and applying the known schedule to produce the load shape. The least expensive device for this application is a hand-held multi-meter capable of measuring power.

Example 4-1: Constant Load Lights

A grocery store has 14 low-pressure sodium lights in its parking lot and the lights are on a dedicated electrical circuit. The lights are operated by a timer and come on at 6 p.m. in the evening and shut off at 7:00 a.m. A spot measurement of the circuit indicates that the power draw when the circuit is on is 10.4 kW. The load shape presented in Figure 4-4 could then be generated. If using this approach it is advisable that you observe the load to ensure that it is actually coming on and going off according to the operating schedule.

Constant Load With Variable Hours

A constant load with variable hours is one where either the operating hours vary daily, seasonally, or are unknown. Examples of constant load with varying hours might include a bank of indoor lights that are manually controlled or an individual office controlled by an occupancy sensor or a motor on an industrial process which has variable operating hours. In this case, the most cost-effective approach is to measure the power being drawn by the loads when they are on and apply the load to an operating schedule determined using a run-time meter. Again, use a held-held meter to

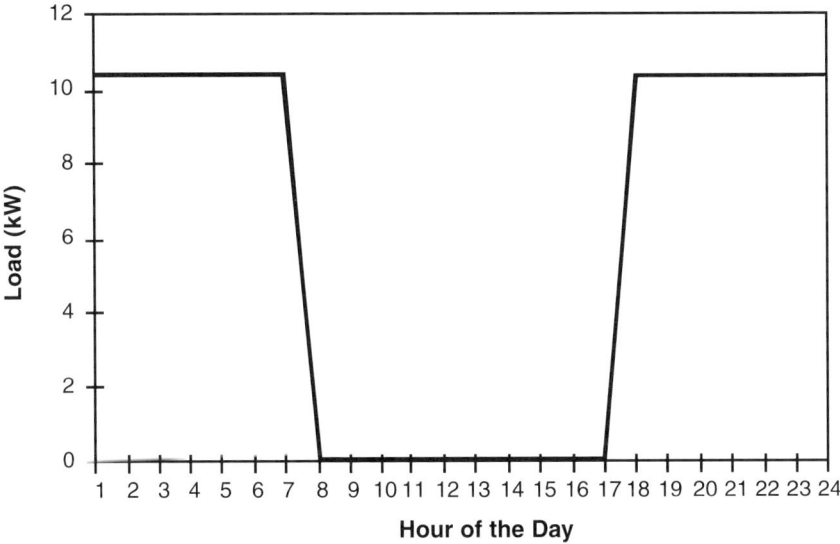

Figure 4-4. Constant load with fixed operating hours.

60 Chapter 4

measure the power load and an inexpensive run-time meter or 1- to 4-channel data logger to measure the hours of operation for the load.

Example 4-2: Variable Hour Lighting

An office building has indoor lights which are controlled by banks of switches. They have eight individual offices in the building with twelve lights on one lighting circuit and four lights on a second circuit. Although you could measure the run-time hours and power for both circuits, an alternate scheme might be to only measure the hours of operation and power consumption on the first circuit, which contains approximately 75 percent of the office space in the building. Although this should be a reasonable approximation of the lighting usage pattern for the building, you would improve your accuracy of the load shape if you obtained data from both circuits. Using the run-time meter you would obtain run-time data as represented by Figure 4-5, which shows the actual times the lights are switched on.

Upon examination of Figure 4-5 it appears that there are some hours when only a portion of the lights are on. Actually, these are periods when the lights are only on for a portion of the hour.

To determine the final load shape, you would multiply the load indicator in Figure 4-5 by the power obtained using the spot meter on the first lighting circuit. Using these data from lighting circuit 1 you could extrapolate the results to the entire office building by

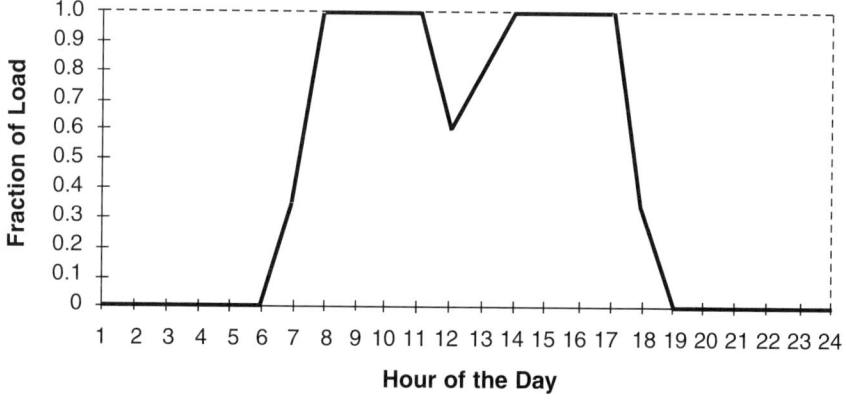

Figure 4-5. Constant load with variable hours.

multiplying the load by the ratio of lights on both circuits by the number of lights on circuit 1, which is 16/12, or 1.33.

Variable Load

A variable load is one where you do not know how much of the load will be on at any point during the day. Examples of variable load might include a suite of offices in an office building where each office is individually controlled by an occupancy sensor or a variable speed motor on an industrial process. In this case it is imperative that you monitor the equipment load in real time. The most cost-effective choice is probably to use a 1- to 4-channel data logger to monitor the power.

Example 4-3: Lights with Occupancy Sensors

An office building has eight offices, each of which has two lights controlled by occupancy sensors. All of the lights in the offices are on a single electrical circuit. Using a 1- to 4-channel data logger to measure the real-time power requirements would obtain run-time data as represented by Figure 4-6.

Metering Plans

In the previous section, some examples of cost-effective methods for obtaining end-use metered data for different load applications

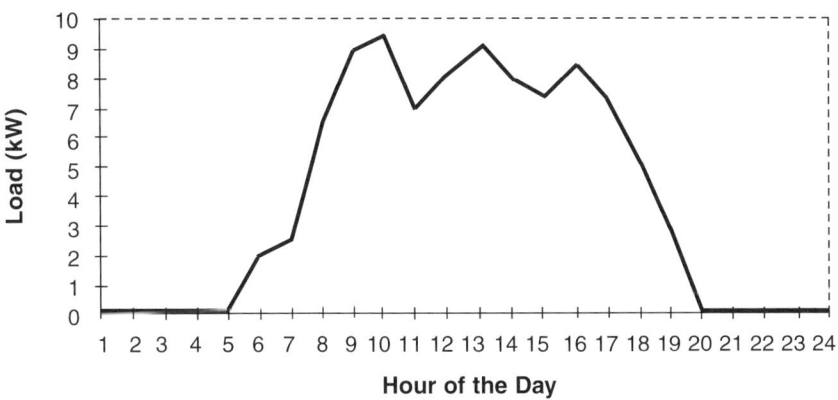

Figure 4-6. Variable load.

were presented. However, before you actually go into an unknown facility and start hooking up metering or monitoring equipment you need to perform a metering study so that your end-use data represents what you thought you were metering. After you perform the metering study you can develop the metering plan. The metering study tells you what energy-using equipment is at the facility and how it is configured, while the metering plan describes how you will hook up your metering equipment for your particular study.

Metering Study

Whether you're gathering whole-building or end-use metered data for a building or meter you need to be sure there is a concise record of what meters serve the site, what circuits are connected to the meters, and what loads are connected to those meters. An example of a three-page generic data collection form for collecting meter and load data is presented as Figure 4-7.

Let's take a closer look at the forms presented in Figure 4-7 to identify the minimum data that is important to collect before you can develop the metering plan. On the first page of the meter data collection form the following information is gathered:

- contact and address information for the facility being metered
- complete list of electric, natural gas, and other fuel meters that serve the facility
- schematic of the facility

On the second page of Figure 4-7, information on which electrical circuits in the building are connected to each electric meter is being collected. For non-electric fuels, the meter may directly feed to the end uses. In addition to the circuits on the meters, the form asks for:

- power characteristics of the circuit (voltage, amperage, etc.)
- description of the circuit which can be either a location served by the circuit (e.g., kitchen) or a description of the end uses served by the circuit (e.g., hot water)

On-Site Meter Data Collection Form

Contact Information:

Company Name: _____
Address: _____
Address: _____
City: _____ State: _____ Zip Code: _____
Contact Name: _____ Title: _____
Phone Number: _____

Meter Information:

	Meter Name/ID	Location	Elec.	Gas	Other
1	_____	_____	☐	☐	☐
2	_____	_____	☐	☐	☐
3	_____	_____	☐	☐	☐
4	_____	_____	☐	☐	☐
5	_____	_____	☐	☐	☐
6	_____	_____	☐	☐	☐
7	_____	_____	☐	☐	☐
8	_____	_____	☐	☐	☐

Building Schematic

Figure 4-7. Generic on-site data collection form for metering—page 1.

On-Site Meter Data Collection Form

Circuit Identification

	Meter ID	Circuit ID	Voltage	Amps	Phase	Description
1						
2						
3						
4						
5						
6						
7						
8						
9						
10						
11						
12						
13						
14						
15						
16						
17						
18						
19						
20						
21						
22						
23						
24						
25						
26						
27						
28						
29						
30						

Figure 4-7. Generic on-site data collection form for metering—page 2.

On-Site Meter Data Collection Form

Device/Appliance Identification

	Circuit ID	Voltage	Amps	Power	hp	Qty.	Name	End Use
1								
2								
3								
4								
5								
6								
7								
8								
9								
10								
11								
12								
13								
14								
15								
16								
17								
18								
19								
20								
21								
22								
23								
24								
25								
26								
27								
28								
29								
30								

Figure 4-7. Generic on-site data collection form for metering—page 3.

If you were collecting data on non-electric circuits, you would substitute the power characteristics with capacity characteristics such as gallons of oil or CFM of natural gas.

On the third page of Figure 4-7, notice how information on the actual energy using devices or end uses in the facility is being collected. Information is being obtained on:

- the circuit that the appliance is connected to
- the power requirements of the appliance
- a description of the appliance
- an indication of which major end use the appliance belongs to (e.g., an oven in a bakery would probably be grouped under the cooking end-use category)

Metering Plans

One of the major concerns when performing metering in a building, whether to collect end-use data or monitor energy and demand savings for a performance contract, is that the metered data be as "clean" as possible. If you are collecting end-use data, clean means that for any given set of end-use data the majority of loads that make up that data belong to a single end use. For example, if you are monitoring the indoor lighting circuit in a facility, you want to ensure that there are not other significant loads on the circuit. If you are monitoring energy or demand savings from an alteration in facility equipment or controls, clean means that you are measuring the amount which energy and demand change in response to those specific facility alterations.

Example 4-4: Determining Cleanness of Metered Data

Let's assume that a bakery has one electric meter which serves two circuits. The data collected in the metering study for the second circuit is presented in Figure 4-8. How clean is the second circuit in the bakery?

First, you need to define what you are trying to meter. In this case, you want to collect metered data on the major end-use loads in the facility. Therefore, you need to look at the power require-

On-Site Meter Data Collection Form

Device/Appliance Identification

	Circuit ID	Voltage	Amps	Power	hp	Qty.	Name	End Use
1	2			100 W		12	ceiling lights	lighting
2	2			150 W		6	ceiling lights	lighting
3	2			800 W		1	computer	miscellaneous
4	2			150 W		40	cafe lights	lighting
5	2			75 W		3	ceiling lights	lighting
6	2			200 W		4	ceiling lights	lighting
7	2			75 W		3	ceiling lights	lighting
8	2			2,000 W		1	oven	cooking
9	2			1,500 W		1	dough mixer	cooking
10								
11								
12								
13								
14								
15								

Figure 4-8. Meter data collection from bakery.

ments for the appliances shown in Figure 4-8 and calculate what percentage of the load belongs to the various end-use categories.

Upon examination of the data in Figure 4-8, the sum of the loads can be calculated for three end uses: lighting, cooking, and miscellaneous. The summations are shown below:

$$\left.\begin{array}{l} \sum \text{lighting} = 9,350 \text{ Watts} \\ \sum \text{cooking} = 3,500 \text{ Watts} \\ \sum \text{miscellaneous} = 800 \text{ Watts} \end{array}\right\} \text{Total} = 13,650 \text{ Watts}$$

Using these data, the percent contributions from each end use are calculated by dividing the total power in each end use by the total power from all the devices on the circuit as shown on the following page:

$$\% \text{ lighting} = \frac{9,350 \text{ Watts}}{13,650 \text{ Watts}} = 0.68 \text{ or } 68\%$$

$$\% \text{ cooking} = \frac{3,500 \text{ Watts}}{13,650 \text{ Watts}} = 0.26 \text{ or } 26\%$$

$$\% \text{ miscellaneous} = \frac{800 \text{ Watts}}{13,650 \text{ Watts}} = 0.06 \text{ or } 6\%$$

Based on these results lighting is the predominant end use on circuit 2, with cooking consuming about one quarter of the circuit's power, and miscellaneous consuming a small portion of the power. Therefore, it is reasonable to conclude that this circuit does not provide clean data for any single end use but provides clean data for the lighting and cooking end uses combined.

The next step in developing a meter plan is to determine what type of metering equipment to use to collect the desired data. Depending on which type of data you are collecting and your budget constraints, select the appropriate metering equipment.

As a final step to preparing the metering plan, be sure to document precisely:

- which equipment or circuits are going to be metered
- where the metering equipment will be located
- what metering equipment will be used

Example 4-5: Develop a Metering Plan

Develop a metering plan for recording the combined end-use load shape data for both the cooking and lighting end uses from Example 4-4. For this project, you might choose to monitor the power used by the equipment on circuit 2 in 15-minute intervals using a relatively inexpensive, 4-channel data logger. This satisfies the basic requirements for this particular project. However, since both of the pieces of cooking equipment (e.g., the dough mixer and the oven) have a relatively constant power draw, a less expensive option is to install run-time meters on those pieces and measure their respective power loads when they are operating at full load.

After the cooking load shapes are obtained, they can be subtracted from the total load on circuit 2 to obtain an end-use estimate for the lights on the circuit. Recall that the miscellaneous

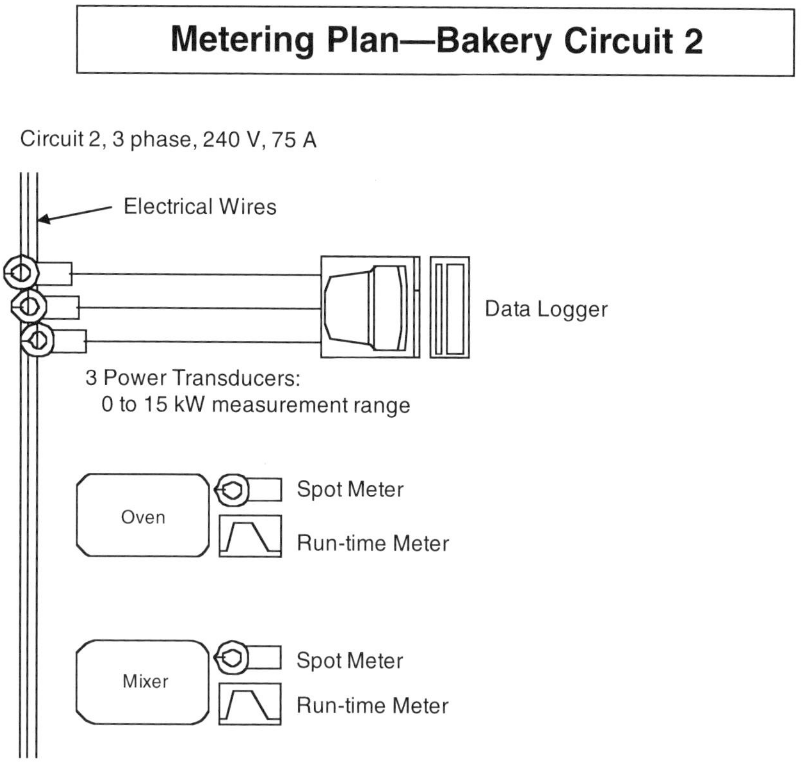

Metering Plan—Bakery Circuit 2

Circuit 2, 3 phase, 240 V, 75 A

- Electrical Wires
- 3 Power Transducers: 0 to 15 kW measurement range
- Data Logger
- Oven: Spot Meter, Run-time Meter
- Mixer: Spot Meter, Run-time Meter

Figure 4-9. Metering plan for bakery circuit 2.

equipment only constitutes about 5 percent of the load on the circuit and could be ignored with little impact on the resulting load estimates. A graphical metering plan for circuit 2 is presented in Figure 4-9. Although, you may not be able to produce graphical representations for each circuit or device you wish to monitor, at a minimum you will need to develop a written metering plan.

Metering Installation

Now that you have a metering plan developed, the next step is to install the meters in the facility. Can you enter the facility and install the meters at your convenience? No, absolutely not. When you install monitoring equipment at a facility implement the following guidelines:

- install equipment after normal business hours
- practice safety first
- keep monitoring equipment out of the way
- test the equipment when it's installed

Install Equipment After Hours

First and foremost, you do not want to inconvenience the facility in any manner that forces it to alter its normal operation. Remember that every facility you install monitoring equipment in is a business and if you disrupt them during their normal business hours they may lose money.

Additionally, try not to disrupt production schedules that continue 24 hours a day. Most production lines will have predetermined break times for routine maintenance of their equipment—if you can install your monitoring equipment during these regularly scheduled break times the facility staff should be most grateful.

Of course, if you are installing monitoring before and after an equipment or efficiency upgrade, or a remodeling, the disruptions have already been scheduled. Just be sure to coordinate the effort with the proper facility personnel.

Practice Safety First

Sometimes the power in a facility cannot be turned off—especially if the facility must operate 24 hours a day. If you absolutely must hook up metering equipment to live wires, buy a pair of insulated rubber gloves as well as shoes with insulated soles.

Second, if you are hooking up monitoring equipment and you are able to turn off the power at the electrical box, be sure to practice safe tagging procedures. These procedures commonly consist of placing a warning tag on the electrical box and perhaps even locking the box with a plastic lock to ensure that nobody turns the power back on while you are installing equipment somewhere else on the electrical circuit. When you have completed installation procedures on the electrical circuit, you should be the one to remove the tag and locks from the electrical box.

Third, if you do not have the expertise to install monitoring equipment sub-contract the work out to someone who does have the experience. Also, you may wish to have a licensed electrician assist you in installing the monitoring equipment at facility sites—otherwise you may be liable for any damages or accidents that occur.

Keep Monitoring Equipment Out of the Way

Following this suggestion is of great assistance to both the facility personnel and yourself. From the facility standpoint, it does not want to have monitoring equipment physically interrupt its normal routine. An example of a physical interruption might be monitoring equipment located in a heavily utilized area where people are forced to alter their traffic pattern to avoid the equipment. From your standpoint, you would like the monitoring equipment to be isolated from facility personnel to minimize the possibility that the equipment is accidentally damaged or disconnected.

If your monitoring equipment becomes disconnected or damaged during a monitoring project, at a minimum, you would have to incur the cost of sending people back out to the facility to reinstall and check the equipment and, in the worst case, you might have to replace the monitoring equipment if it were damaged.

Of course, there have been monitoring projects where facilities have the main electrical panel located in a heavily traveled, narrow space and the only option is to install the monitoring equipment in this location, where the possibility of damage is high and the facility personnel have to be careful to avoid the equipment. But, this should be the exception and not the rule.

Test the Equipment When It's Installed

Finally, if you are going to leave the metering equipment at the facility to record data you need to check and ensure that the equipment is working correctly after it's installed. This will keep you from coming back a couple of weeks later, retrieving the data, finding out the data is bad and going back to the site to figure out what went wrong with the meter installation process.

If the facility has an unused phone line, it is generally cost-effective to put in the capability to dial-up and retrieve the information that is being collected in real time, especially if the facility is located a long distance from where you are or you are collecting data for a long period of time (one month or longer).

So, what do you test while you're still in the field installing the monitoring equipment? You want to ensure that the monitoring equipment is accurately recording the on, off, and part load (if applicable) power, voltage, amperage, or run-time hours of the equipment. For starters use the following guidelines for different types of metering equipment.

Spot Measurements

Before you turn the equipment (or circuit) on, take some readings on the equipment to be sure that you are getting a zero reading. Next, turn the equipment on (to full load if applicable) and spot measure to obtain the full load readings. Since you previously did a metering plan, you should have an estimate of what the maximum possible readings (e.g., voltage, power, etc.) should be based on the nameplate data or engineering estimates of the equipment. Take at least three spot measurements to check for consistency. If you are not obtaining consistent readings, check by using a second spot meter to validate the readings.

Run-Time Meters

Once again, with the equipment (or circuit) turned off, check to ensure that you are obtaining a zero reading indicating the equipment is not running. Then, turn the equipment on and check to see that you are getting a non-zero reading indicating the equipment is running. Double check with another run-time meter if needed.

Data Loggers

Use the same approach as with the spot meters or run-time meters depending on what you are measuring with the data logger. On most data loggers, you should be able to view the data being collected in real-time to validate that it's working correctly.

Equipment Calibration

Be aware that the monitoring equipment you are using may not be properly calibrated. If a capital purchase of monitoring equipment is made, funds need to be available to cover routine maintenance expenses which cannot ethically be passed on to clients. Failure to do this renders the metering equipment effectively useless and the initial capital expenditures are effectively unrecoverable.

When you get monitoring equipment, follow the instructions that come with it and have regularly scheduled calibrations performed on the equipment if applicable. The small investment in time and finances to regularly calibrate your monitoring equipment is well worth the effort.

Data Certification

Once you have performed and checked the installation of the monitoring equipment, you are ready to actually start collecting the end-use data. Whether the data is retrieved manually or through telecommunications equipment, you will need to perform quality control checks on the data to ensure that it's accurately representing what's happening at the facility. What types of things can you check with regard to the data quality? Some of the basic checks that you can perform include range checks, scheduling checks, and consistency checks.

Range Checks

In range checking, you are setting bounds for what you consider to be reasonable data from the equipment (or circuit) that you are monitoring. If you are measuring power and you know that the load can be anywhere from zero or no-load to the full load condition, your bounds would be set at 0 and the full load of the equipment. During the metering installation you should have taken spot measurements of the power the equipment is drawing at full load or you can use an engineering estimate to approximate the full load rating.

If you receive monitoring data that indicates the load is outside of this range, you need to check the data to ensure that it is accurate. It's important to realize that most equipment can operate at a higher load than what is given as "full load" capacity. Most notably, motors and compressors can operate at higher than full load for short periods of time. Light fixtures are also known to draw more than their rated capacity depending upon ambient conditions such as the temperature of the space.

Example 4-6: Performing Range Checks on Data

Perform a range check for monitoring being performed on variable speed motors on an industrial process in a catalyst manufacturing facility. Assume that the facility manager at this particular plant has informed you that five, 20 hp, variable speed motors operate 24 hours per day with a minimum loading of 20 percent and a maximum loading of 90 percent, with an efficiency of 89 percent.

First, let's use engineering estimates to predict the upper and lower bounds on the expected power draw for these motors using the standard equation given as Equation 4-2.

$$Power = \left(\frac{hp \cdot plf}{\eta}\right) \cdot 0.746 \frac{kW}{hp} \qquad \textbf{(Equation 4-2)}$$

where:

  $Power$ = power draw of the motor

  hp = rated horsepower of the motor

  plf = part load factor of the motor (e.g., fractional percent loaded)

  η = motor efficiency

  0.746 = conversion factor from hp to kW

Based on Equation 4-2, and the data given in the assumptions, the load at both the 20 percent and 90 percent load conditions can be calculated as 16.8 kW and 75.4 kW, respectively, for all five motors.

Next, compare the actual recorded power data to the upper and lower bounds to check for inconsistencies in the metered data. A graphical comparison of this can be observed in Figure 4-10. Here we see two potential problem areas with the data.

Before you assume that the data is bad, contact the facility and inform it when and what you see occurring in the monitored data. Perhaps it made changes in the production line or had some other explanation for the data anomalies. Also, check to see if this anomaly is a recurring problem. If it is recurring, it may be a normal occurrence that the facility is unaware of.

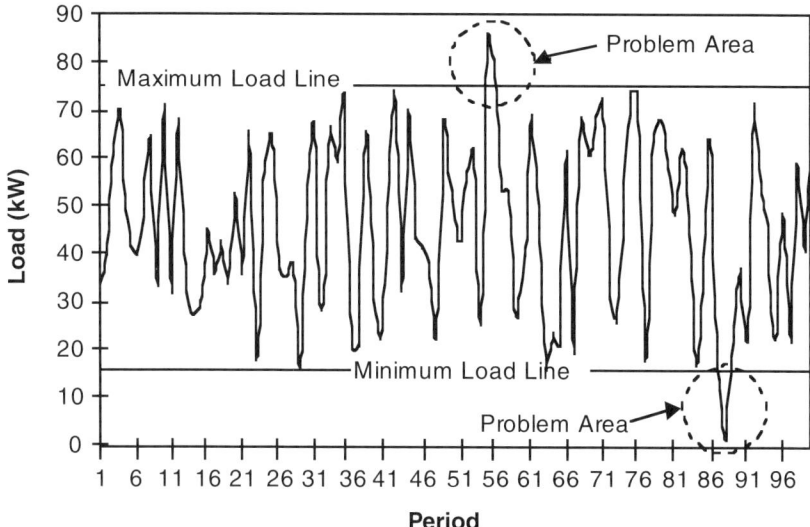

Figure 4-10. Range checking of monitored data.

Schedule Checks

Schedule checks are implemented by comparing the general operating schedule of equipment to the schedule that the monitoring reports for the given equipment. The general operating schedule may either be obtained from facility personnel or observed firsthand.

As with range checks, if you see equipment operating outside of your normal expected time frame, contact the facility and see if it altered its operations in any manner. It is quite common for janitorial staff to come in the building after hours, turn on the lights in all of the offices, and then turn off the light in each office it cleaned. This way, it is able to keep track of which offices still need to be cleaned. However, very few people in the company are generally aware of this practice and will tend to claim that almost all of the lights are turned off at night.

Consistency Checks

Another mechanism that can be used to ascertain whether a data anomaly is bad data or good data is to check to see if you can spot the same anomaly repeating itself at other times. Chances are that if an anomaly repeats itself it's actually good data but the facility personnel simply are not aware of the operating practice which you are observing.

Reference

Bowman, M., and Goldberg, M. L., "Turning Lead into Gold: Trends in Metering for Studies of Load and Load Impacts," Proceedings from the ACEEE 1994 Summer Study on Energy Efficiency in Buildings, Vol. 2, p. 2.31, ACEEE, 1994.

5
Data Collection

What Role Does Data Collection Play?

The purpose of data collection is to find out about what kinds of equipment your customers currently use, would like to use in the future, or find out their attitudes concerning equipment purchases, Demand-Side Management (DSM) programs, your utility, or other issues. Collectively, all of these data collection aspects are referred to as market research. This chapter will focus on how to obtain information about the equipment and operating schedules in a facility, since these data can be used to produce end-use load shapes.

There are several different kinds of data collection activities that can be performed, including:

- mail surveys
- phone surveys
- on-site data collection

Survey Design

Regardless of which type of data collection is performed, it is important that the survey be properly designed to (1) provide a clear understanding of what is being asked for, (2) be unbiased in obtaining the data, and (3) ensure that the data collected will support whatever analysis will be performed using the data.

Mail Surveys

A mail survey is probably the least expensive way to obtain data about your utilities' customers. If you produce a simple one-page mail survey you can probably arrange to have it sent out to customers for about $5 per survey, including labor, assuming a labor cost of $40 per hour. Some of the downsides of performing a mail survey are a relatively low response rate from customers and concerns about data accuracy.

Response Rate

The response rate is simply the ratio of how many surveys you get back divided by how many surveys you sent out. Typical response rates for mail surveys are around 20 percent to 30 percent. Some methods that have been used to increase a survey response rate are to enclose a dollar bill or promise a free gift if the respondent returns the survey. I personally have received both of these enticements and in both cases was willing to fill out and return the survey.

Of course, utilities are in the unique position of being able to offer energy- and cost-savings incentives to their customers, including such things as a free compact fluorescent light bulb or a $2 reduction in the next months' energy bill. Another method that can be used to increase response rates is to telephone the potential respondents ahead of time to determine if they are willing to take the time to fill out a survey.

Accuracy Concerns

The decrease in accuracy can occur for several reasons, including:

- the respondent does not understand the question
- the respondent misunderstands what equipment he/she has

You are probably aware of the acronym KISS which stands for Keep It Simple Stupid. This is never so true as when you are sending people a survey to fill out. If you send them a complicated questionnaire, some respondents will invariably have difficulty with it.

Primary Heating System *Secondary Heating System*

1. What type of primary heating system do you have in your home?
   - ☐ Furnace
   - ☐ Baseboard
   - ☐ Heat Pump
   - ☐ Wood Pellet Stove
   - ☐ Boiler
   - ☐ None

2. What fuel does your primary heating system use?
   - ☐ Electric
   - ☐ Oil
   - ☐ Gas
   - ☐ Wood/Pellets

3. What fraction of your total heat use is supplied by the primary heating system?
   ☐☐☐ %

4. What is the efficiency of your primary heating system?
   ☐☐☐ %

1. What type of secondary heating system do you have in your home?
   - ☐ Furnace
   - ☐ Baseboard
   - ☐ Heat Pump
   - ☐ Wood Pellet Stove
   - ☐ Boiler
   - ☐ None

2. What fuel does your secondary heating system use?
   - ☐ Electric
   - ☐ Oil
   - ☐ Gas
   - ☐ Wood/Pellets

3. What fraction of your total heat use is supplied by the secondary heating system?
   ☐☐☐ %

4. What is the efficiency of your secondary heating system?
   ☐☐☐ %

Figure 5-1. Potentially "complicated" heating questions.

Example 5-1: Survey Development

Assume you are developing a series of questions that concern the heating equipment and its operation in the home as shown in Figure 5-1. While an energy consultant may find these questions very understandable, some of them may be confusing to some respondents. For example, how does a respondent determine the fraction of heating that is supplied by his/her primary and secondary heating sources. Some respondents may also be confused by the terms "primary" and "secondary" themselves.

An example of a questionnaire that contains a less complicated questions about people's heating systems is shown as Figure 5-2. It's important to realize that when you ask less complicated questions, either by eliminating some questions or rewording them, you may potentially sacrifice the accuracy of the data collected. For example, if you don't ask about the efficiency of heating equipment directly, you would have to make assumptions about efficiencies.

Main Heating System

```
1. What type of heating system do you
have in your home?
    ☐ Furnace          ☐ Baseboard
    ☐ Heat Pump        ☐ Wood Pellet Stove
    ☐ Boiler           ☐ None

2. What fuel does your heating system use?
    ☐ Electric         ☐ Oil
    ☐ Gas              ☐ Wood/Pellets

3. How often do you also use wood/pellets
to heat your home?
    ☐ Never            ☐ Sometimes
                         (once per week)
    ☐ Rarely           ☐ Always
      (once per month)
```

Figure 5-2. Less "complicated" heating questionnaire.

Self-Contained Package

In a mail survey, you need to send a self-contained package to the respondent. This means that in addition to the actual survey, you need to send detailed instructions with examples so that the respondent understands precisely what you are asking for. This is especially critical if you are asking technical questions of non-technical people.

It is helpful to the respondents if you provide a toll-free telephone number so that they can contact somebody if they have questions. If you are having people fill out a scan sheet that requires that a special pencil (such as a No. 2) be used to fill out the form—send them a No. 2 pencil. People are busy and are rarely going to run to the store, buy a No. 2 pencil, run back home to fill out the form, and put it back in the mail to you. And finally, don't forget to have a self-addressed, stamped envelope so the respondent can send the answers back for processing. It doesn't do you any good if 40 percent of your sample fills out the form but doesn't send it back because they don't have a stamp or aren't willing to spend 32 cents on your survey (and who can blame them). Remember that the respondents are doing you a service so make it as easy as possible for them to participate in the survey.

Premise Definition

Issues about the unit of analysis can also affect your responses. Recognize that residential respondents are going to answer questions about their household or dwelling unit assuming that's the unit of analysis, and that commercial and industrial customers probably will answer questions assuming that their place of business is the unit of analysis. If you assumed that the unit of analysis is the electric or gas meter and have performed a sample design based on those assumptions, your results may not actually be representative of the utility population.

Phone Surveys

Phone surveys fall between mail surveys and on-site data collection in terms of cost to perform the survey. The main factors that affect the cost of phone surveys are how long it takes to complete the average survey (this includes the time it takes to contact and talk to both people who do participate and those who decline to participate). For example, if you have someone making calls who costs $60 per hour and he/she is able to conduct 10 surveys in an eight-hour day, the average cost of the surveys is $48 per respondent.

Response Rate

The response rate for performing phone surveys varies greatly. As phone fraud increases, people are increasingly unwilling to provide information over the telephone, especially demographic information such as age, race, salary, etc. It's important that people are given a phone number at the sponsoring utility that they can call to verify that you are either a utility employee or providing a service on behalf of the utility.

Accuracy

Data collected from phone surveys should have a higher accuracy than data sent in by mail surveys. The primary reason is that the person asking the questions is available to answer any questions and interpret the answers of the survey respondent. It's important that the people asking the questions are provided with a script as well as being trained in how to interpret responses. Additionally, it's critical that people involved in collecting the data over the telephone or on-site are aware of how the data will be utilized in the analysis. At a minimum, people should be informed about what questions are the most important to obtain.

Script Questionnaire

In telephone surveys, it is important that the telephone personnel are professional, obtain the best data possible depending upon how the respondent answers the questions, and, hopefully, are friendly and cordial. Figure 5-3 presents part of an example script questionnaire that prompts the person reading the questions to ask different questions depending upon the respondent's answers.

On-Site Data Collection

On-site data collection is by far the most expensive and the most accurate method of collecting information on energy using facilities at a site. The cost for performing a survey on a small commercial building is in the range of $100 to $200, but can increase to several thousand dollars for large industrial facilities and depending upon the quantity of the data gathered at the site.

Response Rate

The response rate for performing on-site surveys appears to be quite good, especially if you are willing to provide the survey participants a breakdown of the energy-using equipment at their site along with recommendations for cost and efficiency improvements. It appears that small- to medium-sized facilities that do not

---

Introduction

Hello, my name is _____ and I am calling on behalf of Generic Utility Power to obtain information on your future plans for purchasing energy efficient equipment. Do you have 10 minutes to answer some of our questions?
If YES go to **Questionnaire,** if NO go to **Schedule Appointment.**

Schedule Appointment

1. What time would be convenient for you?
If they decline _thank them_ and indicate they are unwilling to participate.

Name: _____

Date: _____

Time: _____

Questionnaire

1. What types of energy purchases would you consider in the next year? Please respond with a yes or no to each of the following items:

    a. energy efficient motors   ___Yes   ___No
    b. compact fluorescent lights   ___Yes   ___No

Figure 5-3. Script questionnaire.

have in-house energy management personnel are the most eager to participate in an on-site survey study.

Accuracy Concerns

The accuracy of data collected on-site by professional data collection personnel should be much higher than data collected by either mail or phone surveys. In addition, going on-site allows more detailed information to be collected including the operating schedules of individual pieces of equipment. The accuracy of data collected from on-site data collection depends upon:

- how well the auditors are trained in equipment identification
- how accurately the auditors can determine the premise
- how large the facility is
- how much time the auditor has on site

Scheduling Surveys

When performing on-site surveys you must schedule an appointment with the facility ahead of time, as opposed to phone surveys where you can make a "cold" call without previously contacting the facility. It is helpful if the utility can send letters to potential survey participants ahead of time to let them know that a scheduler may call them to make an appointment.

The scheduler needs to be a person who is able to quickly identify the following information:

- is the facility willing to have an on-site study performed?
- who is the correct contact person at the facility to schedule with?
- basic site information
- appointment date, time

Is the Facility Willing?

In general, most facilities are willing to let utility personnel, or those people performing work for a utility, come in and inventory and study their energy-using equipment, especially if they receive recommendations on how to lower their energy costs or improve the efficiency of their equipment.

Every project will probably encounter potential participants who refuse to participate due to concerns that the data will not be kept confidential and get in the hands of their competitors. Examples of these customers might include defense contractors or semi-conductor manufacturers. To help minimize these concerns, the utility can promise that no information will be released on individual customers in the study. Only "typical" information such as the average number of motors per industrial customer or "aggregate" information such as the total number of motors among all of their industrial customers will be released.

It's important to realize that even with these constraints on the release of information some customers may be unwilling to participate. If customers don't want to participate you should respect their wishes, even if they are contractually bound to participate. Be especially aware that as a deregulated energy market emerges a customer may not be willing to tolerate what he/she perceives as constant "nagging" by his/her energy provider and he/she may switch providers.

Correct Contact Person

Once the facility has agreed to let you come in and perform a survey of its equipment, the next most important factor in performing a successful audit is to ensure that you have contacted the correct person at the facility and scheduled an appointment with him/her. Who is the correct contact person? Simply, the person who knows the most about the energy-using equipment and operation of that equipment in the facility. It's important to understand that the correct person may be the owner of the company or he/she may be on the bottom rung of the company ladder. In either case, he/she is the most important person to you—don't be afraid to let him/her know that and ask if you can contact him/her at a later date if you have any questions about the facility.

Once again, if you are asking questions about how the facility makes purchasing decisions on energy equipment the correct contact could be the president, the chief financial officer, the facility manager, the head of the production line, or somebody else. Realize that if you are asking market research questions in addition to collecting on-site data you will probably have multiple contacts at a single site and need to coordinate with them to split up your time as effectively as possible.

- Along with the names of the contacts, it is good practice to collect the contacts' title, phone number, and fax number.

Basic Information

The third important item when scheduling an appointment is to be sure that you obtain basic information about the facility, including:

- the square footage of the facility
- annual energy use/peak demand

These two pieces of information help the scheduler make an educated guess as to how much time will be required for the auditor to perform the survey at the site. Small sites (up to 10,000 square feet) can probably be audited in half a day, medium sites (between 10,000 and 20,000 square feet) can probably be audited in one day, and larger sites can take up to several days to audit. However, it's important to know how long it takes to collect data using your particular survey instrument. If you are using a brand new data collection survey instrument, or one that has significant modifications from a previous study, it is worth the effort to perform some test surveys to check both the time it takes to audit different facilities and the suitability of the data for future analysis.

The scheduler should have obtained the annual energy use and peak demand information from the utility ahead of time. You must realize that the facility may have better information on the energy use at its site since the utility's records usually deal with individual meters and the utility may not have records on which or how many meters are used at a single site, especially if the energy "bill" is sent to a different address for payment.

When a scheduler is arranging appointments, he/she wants to maximize the efficiency of the auditors which means keeping them as busy as possible during the workday while still allowing them sufficient time to collect the on-site information. Additionally, if audits are being performed at a remote location maximize the number of audits that can be performed at that location. If you have an auditor traveling to a remote site for a single audit and you are paying for airfare or hotel, the cost for that audit can easily be double or triple the cost of sending an auditor to a local site. You minimize the additional travel and lodging costs if you spread the cost over several audits.

Schedule Appointment

As a final step you need to actually schedule the on-site data collection effort. It's important to be sure that all the necessary contacts will be at the site and available for their respective portion of the audit. This does not mean that they necessarily need to accompany the auditor around the facility, but they can't be in meetings or otherwise unavailable the entire time the auditor is on-site in case the auditor has questions.

Training Auditors

While all of the previous scheduling steps help to ensure a quality audit, in the end the data collected will only be as good as the auditor sent to the site. Auditors need to be familiar with several aspects of data collection including equipment identification and premise identification.

Equipment Identification

Equipment identification is the ability to be able to go to a facility, have someone show you the equipment, and you be able to identify specifically what type of equipment it is. For example, if you are shown parts of the air conditioning system in a facility, you should be able to ascertain which piece of the air conditioning system you are looking at (e.g., the condenser, cooling unit, water tower, etc.) and what kind of equipment it is (e.g., a centrifugal chiller, absorption chiller, fan coil unit, etc.).

Although some of this material can be presented and learned in a classroom atmosphere, the best way to learn it is to actually go into the field with an experienced auditor and get hands-on experience. Make sure that your auditing staff is sufficiently trained ahead of time or is able to go in the field with more experienced auditors. At most sites, the facility manager will have a great deal of knowledge about the site and be able (and hopefully willing) to help the auditor with equipment identification if it's necessary.

One tool that the auditor can use while on-site to help identify equipment is to take a Polaroid snapshot of the equipment and present it to an experienced auditor at a later date for identification. Be sensitive that some sites will not allow **any** photographs to be taken at their site—it may either be illegal, as is the case in some high-security areas, or they may simply be worried about confidentiality. In either event, do not enter an industrial facility and start taking snapshots of the production line or manufacturing processes without permission.

Premise Definition

The premise definition can potentially be one of the most difficult issues that an on-site auditor can face, especially if the premise is defined as the end uses served by a whole-building load meter or end-use meters at the site. If you have a sample design for your study then the sample design already defined the premise definition.

To refresh your memory, in a sample design you need to define what the "unit" is that you will be sampling from. The unit could be the equipment served by a particular meter, a building unit (such as an office building or a grocery store), a single-family dwelling, etc. However, in order to do the sample design you need to have information on energy use (or another design variable) for all of the units in your population. In the utility industry the most common available unit has traditionally been the billing meter.

If the utility has already performed a sample design and implemented whole-building metering on a set of buildings, chances are that it has used the billing meter as the sampling unit. Therefore, when you perform an on-site audit on a sampling unit you need to use the same definition as the sample design or the data you collect will not match the data that is actually on the sampling unit. This process of determining what the sampling unit is known as defining the premise and it needs to be done for each building that you audit. Of course if the utility has defined the sampling unit as the entire building your job as an auditor gets much easier.

Assuming that the sampling unit is a single meter, some of the types of situations that you can encounter while trying to define the premise are:

- single meter, single business
- single meter, multiple businesses
- multiple meters, single business
- multiple meters, multiple businesses

Single Meter, Single Business

This is the easiest case to handle when performing data collection at a facility. An example of this would be a gas station served by a single electric meter as seen in Figure 5-4. In this case, the business is served by only one meter. Therefore, when you go on site to collect information on the equipment connected to the meter, you collect information on everything in Roy's Garage. Even if the meter is connected to several buildings that belong to the same business you ideally still would collect information from all of those buildings.

One exception to this may be extremely large sites, such as a college or university campus, which have a single meter. In this case it is cost prohibitive to survey each and every building on campus so you might survey a sample of the buildings on campus and aggregate them up to represent the entire campus.

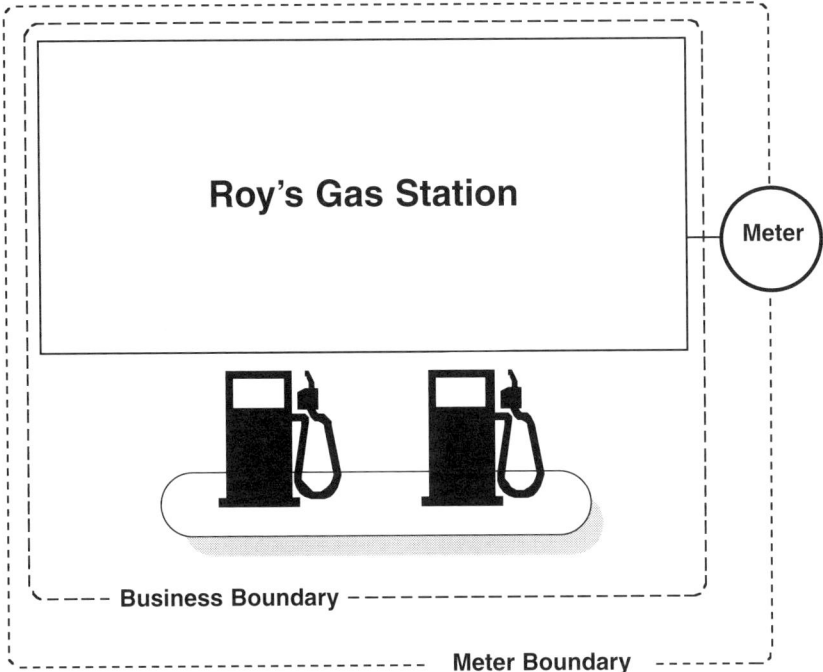

Figure 5-4. Single business, single meter.

Of course, you also need to consider whether or not the site is representative of your population. You have several options when dealing with this situation:

- If you only have one large site of this type you may wish to exclude it from your study and redo the sample design for the remaining buildings in the population.
- You could decide to survey the large site with certainty and redo the sample design for the remaining buildings in the population.
- You could go ahead and survey the site realizing that since it's not feasible to survey the entire thing it may decrease the accuracy of your population estimates.

Single Meter, Multiple Businesses

The next case we will examine is where you have multiple businesses served by a single meter. An example of this would be two

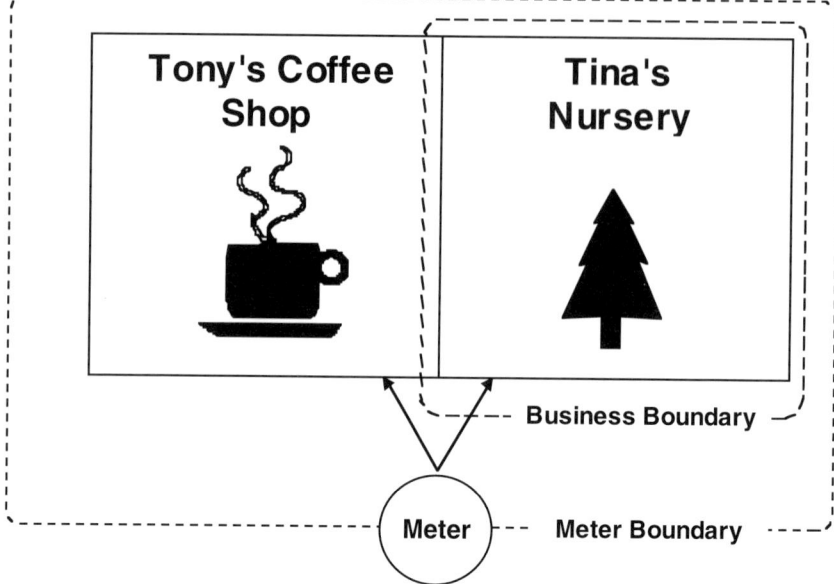

Figure 5-5. Single meter, multiple businesses.

stores in a shopping mall which share the same meter as shown in Figure 5-5. In this case, the meter serves both businesses. Therefore, when you go on site to collect information on the equipment connected to the meter, you should collect information from both Tony's Coffee Shop and Tina's Nursery.

Of course there are a couple of problems that arise with this situation. The first problem is that you only schedule to do the audit with one business (assume Tina's Nursery). Now you are forced to make a cold call on Tony's Coffee Shop to see if they are willing to let you collect data on equipment at their store and/or they have the time to assist you.

Multiple Meters, Single Business

The third case we will look at is where you have a single business that is served by multiple meters. An example of this would be an automobile dealership (Frank's Used Cars) which over time has added a new body shop and separate meters for the body shop and the car lot as shown in Figure 5-6. In this case the business boundary includes both the body shop and the car lot. However, the meter boundary for meter A includes only the body shop.

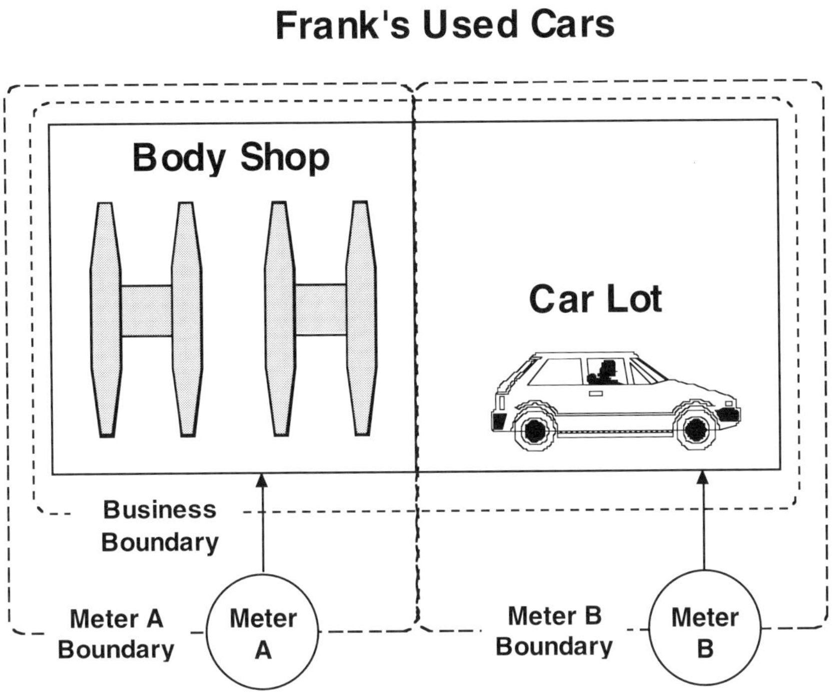

Figure 5-6. Multiple meters, single business.

The problems with this case lie in two areas: first, you need to determine which meter to sample if the sampling unit is a single meter. This can usually be figured out by examining the meter numbers to determine which meter was chosen as a sample unit. After you figure out the correct meter, the second problem is to identify which equipment is served by that meter.

In the case of Figure 5-6, this is fairly straightforward since each meter serves a designated area. In other situations you might have a single building served by two or more meters. In this case you need to rely on the facility manager's knowledge of the electrical layout as well as examining the electrical boxes to see if major end uses are identified at the source.

Multiple Meters, Multiple Business

The last case we will examine is where you have multiple businesses that are served by multiple meters. An example of this

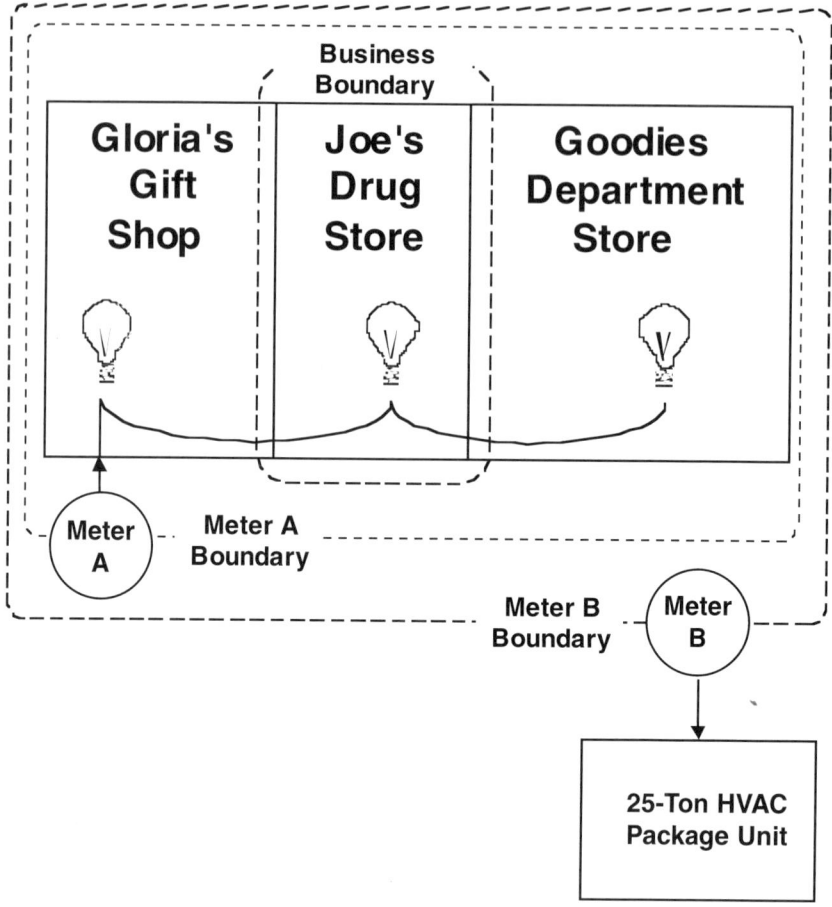

Figure 5-7. Multiple meters, multiple businesses.

would be in a strip mall that has dedicated meters for appliances for each business but has a single meter for the air conditioning and a single meter for the lighting which serves all of the businesses in the mall as depicted in Figure 5-7. Assuming that you had contacted Joe's Drug Store for the survey you would have a business premise as shown in Figure 5-7. However, each meter has a boundary that completely encompasses all three of the stores.

First you need to figure out which meter you need to survey. If the meter is the one with the lighting and appliances you would need to try to collect the appropriate information from each of the businesses. This has the same problems as the single meter, multiple business case. If, however, the meter you need to survey is meter B your job potentially just got easier. If you're working on a study that is just interested in an equipment inventory of the population, you can simply collect the HVAC equipment information from meter B and be finished. If, however, you are collecting information for a building simulation model where you would model the HVAC usage, you would have to collect complete end-use data from all of the businesses on the meter.

Additionally, if in the multiple business cases you need to collect attitudinal information from the businesses you are surveying, what do you do? Only collect attitudinal information from Joe's Drug Store, but if you're collecting end-use data from all three stores this may not be your desired approach.

It's important to realize that before the study is begun decisions need to be made on how these various cases are going to be handled, and then the auditors need to be trained how to handle these cases in the field. Realize that if you schedule your auditors too heavily they may not have time to finish data collection in the unexpected cases where they end up surveying more square footage than was indicated in the original contact information.

Touring the Facility

When you first arrive on-site to audit a facility you may not want to just rush in and start figuring out the correct premise. One method that tends to work well for auditors is to use a three-step approach for gathering on-site information:

1. Take a tour with the facility manager to view the major energy-using components of the facility.
2. Go off on your own to actually do the data collection effort.
3. Reconvene with the facility manager near the end of your time at the site to clear up any questions you may have.

Take A Tour

In this first phase of the data collection effort you should take a tour with the facility manager in order to get the layout of the facility. You are primarily interested in the energy-using equipment in each area of the facility. For example, in industrial sites you would want to be shown the production process and have people identify large motors, kilns, presses, and so on. During this stage you might take notes on what energy-using equipment is located in each area. The facility manager may already have developed a list of energy-using equipment at the site—if so, use it as your starting point rather than starting from scratch.

Collect Data

After the tour stage, you should be able to wander off on your own and collect specific data from the equipment at the site. The type of data that you collect varies by the type of equipment you are examining. Table 5-1 presents a list of some of the data that you may wish to collect for different types of equipment. This list may be too detailed or not detailed enough depending upon your needs. If you are going to perform a building simulation or engineering model of the building you need to err on the side of collecting too much information. However, if you are simply looking at market share for major end-use groups (cooking, heating, etc.) by fuel type you don't need to collect this much information. Table 5-2 presents some general characteristics that you might need to gather at the site.

Final Meeting

When you are getting ready to finish the data collection effort be sure to talk to the facility manager if you have any more questions or need additional data. Also be sure to thank him/her for his/her assistance and ask if you may call him/her if any questions arise during the analysis.

Quality Control Checks

After the survey is completed, one good quality control check that the auditor can perform is a quick estimate of energy use based on the data collected from the on-site survey and compare the calculated energy use to the energy bills for the facility.

Other simple quality control checks include calculating the square footage of area cooled per ton of air conditioning (ft^2/ton) and comparing those to typical values for different types of buildings. Alternatively, you may calculate the Energy Use Index (EUI) for each end use (kWh/ft^2 or $Mbtu/ft^2$) and compare those to typical values. Table 5-3 presents typical ft^2/ton values for different types of buildings (American Society of Heating, Refrigerating and Air Conditioning Engineers, 1987).

Table 5-1. Important equipment characteristics.

Equipment Type	Characteristics	Units/Notes
Air conditioning	Cooling capacity	Tons
	Efficiency	EER, SEER
	Nameplate rating	kW, volts, amps, BTU
	Area cooled	Square feet
	Type of cooling equipment	Chiller, DX unit, heat pump, room unit, etc.
Space heating	Heating capacity	BTU, kW
	Efficiency	AFUE, COP, efficiency
	Nameplate rating	kW, volts, amps, BTU
	Area heated	Square feet
	Type of heating equipment	Boiler, heat pump, furnace, etc.
	Fuel type	Electric, gas, oil, etc.
Ventilating fans	Size	Horsepower
	Nameplate rating	kW, volt, amp
	Part load factor	Percent
	Type of ventilation system	Constant volume, VAV, etc.
	Air moved	CFM
Indoor lighting	Type of fixtures	T8, compact fluorescent, etc.
	Type of ballast	Magnetic, electronic, etc.
	Power rating	Watts
	Operating schedule	Hours on diversity factor
	Area served	Square feet
Outdoor lighting	Type of fixtures	High pressure sodium, mercury vapor, etc.
	Power rating	Watts
	Operating schedule	Time on, time off
	Control mechanism	Timer, photovoltaics, etc.
	Area served	Square feet

(cont'd)

Table 5-1. Important equipment characteristics *(cont'd)*.

Equipment Type	Characteristics	Units/Notes
Hot water	Size	Gallons
	Type	Tank, boiler, Instantaneous, etc.
	Temperature	°F or °C
	Rating	kW, BTU, Etc.
	Fuel type	Electric, gas, oil, etc.
	External insulation	None, blanket
Cooking	Type	Oven, coffee pot, etc.
	Rating	kW, BTU
	Fuel type	Electric, gas, oil
	Breakfast, lunch and dinner schedule	Times used How heavily used
	Ventilation type	None, hooded
	Ventilation rating	CFM, Hp, etc.
	Area served	Square feet
Refrigeration	Case type	Coffin, refrigeration case, cooler, etc.
	Compressor size	Hp
	Case temperature	°F or °C
	Case size	Square feet or length
	Defrost type	Ambient, electric resistance, none, etc.
	Area served	Square feet
Miscellaneous/ process equipment	Type	Name process type
	Rating	kW, BTU, etc.
	Schedule	Hours on diversity factor
	Cycling schedule	15 min. on, 20 Min. off, Etc.
	Fuel type	Electric, gas, oil, etc.
	Area	Square feet

Table 5-2. General characteristics.

Information	Characteristic	Notes/Units
Occupancy	People	Number
	Schedule	Hours occupied
Building shell	Wall area	Square feet
	Roof area	Square feet
	Window/glazing area	Square feet
	Insulation levels	R-11, R-19, etc.
	U values	BTU/hr-ft^2-°F
	Square footage	Square feet
Thermostat schedules	Cooling setpoint	°F or °C
	Cooling schedule	Hours on
	Heating setpoint	°F or °C
	Heating schedule	Hours on
	Setback temperatures	°F or °C

Table 5-3. Typical square foot per ton values.

	Refrigeration (Ft^2/Ton)		
Classifications	Low	Average	High
Apartment, high rise	450	400	300
Schools, colleges, universities	240	185	150
Industrial, assembly	240	150	90
Industrial, light manufacturing	200	150	100
Industrial, heavy manufacturing	100	80	60
Office buildings	360	280	190
Residential, large	600	500	380
Residential, medium	700	550	400
Department stores	240	160	105
Drug stores	180	135	110
Shopping malls	365	230	160
Hotels, motels	350	300	220
Libraries, museums	340	280	200
Hospitals, patient rooms	275	220	165
Hospitals, public areas	175	140	110

Customer Tracking

If you are doing a study that is using a sample design, it is enormously helpful if you set up a customer tracking system. A customer tracking system should be capable of providing the following information for each sector and strata of the sample design:

- number of sites to choose from
- number of sites in sample design
- number of sites contacted
- number of sites surveyed to date
- number of sites scheduled to be surveyed in the future
- number of sites declining
- number of sites reclassified
- number of sites still available to schedule
- number of sites remaining to meet goal

Number of Sites to Choose From

The number of sites to choose from could be the number of units in the population for the given sector and strata. However, if you are doing a study of a subgroup of the population, such as participants in a DSM program or the customers that have whole-building load data, the number of sites you have to choose from could be much smaller than your population.

Number of Sites in Sample Design

This is the number of units determined in the sample design. This should be a fixed number at the beginning of the data collection process.

Number of Sites Surveyed

This is the number of where data collection has already been performed. Realize that to keep accurate tabs on this value, the scheduler must inform you whenever data collection has been completed at a facility.

Number of Sites Scheduled

This is the number of sites that have been scheduled to be audited but have not yet been completed by an auditor.

Number of Sites Declining

This is the number of sites that have refused to participate in the data collection study.

Number of Sites Reclassified

This is the number of sites that have been reclassified from one sector and/or strata to another sector or strata. When the sample design is based on customer bills from the utility records, there are two common problems that arise.

The first problem is that the customer may have a different energy use than indicated by the utility company—most commonly a customer has multiple meters and the utility did not identify all of the meters that belong to that customer and hence they were initially classified in the wrong strata.

The second problem is that utility customers are commonly classified by SIC codes which may not properly reflect the actual type of facility. For example, a convenience store may be classified as a gas station since it has gas pumps, which may be in the service, retail, or another sector depending upon the sample design. However, when the auditor goes to the site he/she ascertains that the facility is much more representative of a grocery store and therefore reclassifies it from the original sector to the grocery sector. If you change the sector or strata of a site, you also need to change the number of sites available, the number of sites contacted, and the number of sites surveyed in both the original and new categories.

Example 5-2: Strata Reclassification

The utility records indicate that Phil's Grocery has a single electrical meter with an annual energy usage of 25,000 kWh. The sample design defined the grocery sector as having three strata defined below:

- strata 1: <10,000 kWh
- strata 2: 10,000 to 40,000 kWh
- strata 3: more than 40,000 kWh

Upon inspection of the site, the auditor determines that the site is actually served by two meters with an annual energy usage of 65,000 kWh. The auditor would then reclassify the site from strata 2 to strata 3.

Number of Sites Still Available

The number of sites still available is simply calculated by subtracting the number of sites already contacted from the number of sites available.

Number of Sites Remaining

The number of sites remaining is the number of sites you still need to survey to meet your sample design target. This is calculated by subtracting the number of sites surveyed and the number of sites scheduled from the number of sites in the sample design.

Reference

"Pocket Handbook for Air Conditioning, Heating, Ventilation, and Refrigeration," ASHRAE, American Society of Heating, Refrigerating and Air Conditioning Engineers, 1987.

6
Metered Data Analysis

It's Time to Get Real

Metered data analysis is one of the issues that is at the center of load shape development, due largely to the perception that metered data is the marker, or gold standard, against which all other end-use load shape estimation methods can be judged. While the previous chapters in this book have focused on providing you background material, this book will now begin to make a transition from discussing broadbased background material to showing you how to put all of the pieces together and perform some real analysis. Before we get to the analysis stage however, we will present more material on defining the premise. The emphasis in this chapter will be on how to use end-use metered data as opposed to how to collect end-use data, as was previously presented in Chapter 5.

Defining the Premise

As stated in Chapter 5, defining the premise and then collecting data on the premise can be an extremely difficult challenge for on-site auditors, especially if the premise is defined as the end uses served by a whole-building load meter or end-use meters at the site. As in Chapter 5, the premise definition should be performed during the sample design stage—not at the analysis stage of a project.

When performing analysis on metered data from facilities there are two broad categories of data that you will probably have to deal with: whole-building load research data or end-use metered data. When utilizing whole-building load research data in your analysis, there are several types of situations that you can run into, including:

- single meter, single business
- single meter, multiple businesses
- multiple meters, single business
- multiple meters, multiple businesses

However, when a utility (or another party) has monitored end-use data at a particular site the situations become more complicated. Some of the situations that arise include:

- single site, all end use metered
- single site, some end uses metered
- single site, some end uses combined
- single site, only whole-building metered

Whole-Building Metered Data

Whole-building metered data is also commonly referred to as load research data due to its primary use by utilities as a research tool. When you are analyzing whole-building metered data you need to ensure that you understand which end uses are connected to the meter. Hopefully this data has been provided by the utility, another party, a survey, or a metering plan. You may recall from Chapter 5 that the best source of information on what was metered at the site is the metering plan if it's available.

Single Load Research Meter, Single Business

This is the easiest whole-building case to handle when analyzing metered data. An example of this would be a convenience store (Hasty Mart) with a single whole-building load research meter as shown in Figure 6-1. At a minimum you will want to know which end uses are connected to the load research meter and, depending upon the analysis technique that you will use, an approximation of the connected load for each end use as well as an idea of the operating schedules for each end use.

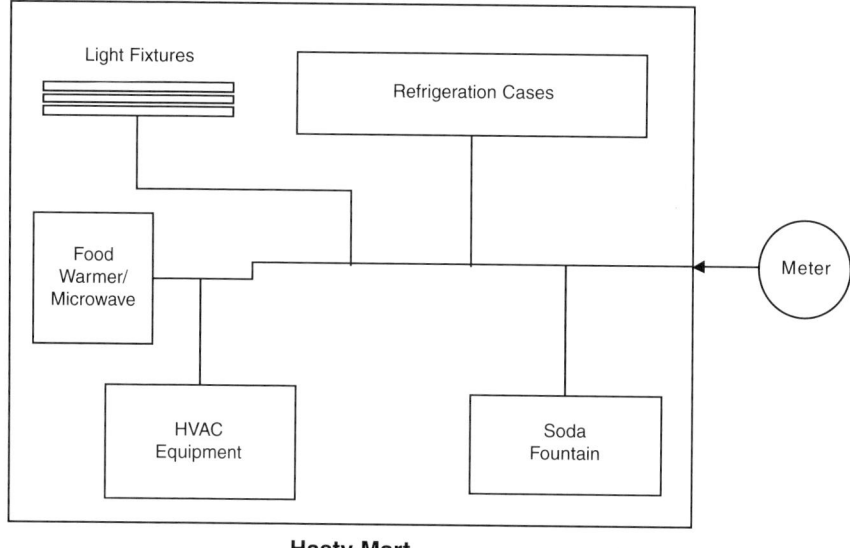

Figure 6-1. Single load research meter, single business.

Single Meter, Multiple Businesses

The next case we will examine is where you have multiple businesses served by a single meter. An example of this would be two stores in a shopping mall which share the same meter as shown in Figure 6-2. In this case, the meter serves both businesses. Assuming that we are still dealing with Tony's Coffee Shop and Tina's Nursery, as in Chapter 5, we need to determine which end uses are connected to the utility meter. In addition, we really need to know which end uses are in each business because this information may be needed depending on whether we chose to analyze these sites as individual businesses or combine them into a new combined site.

Figure 6-2 shows some of the end uses serve both facilities while others are unique to one facility or the other. End uses which are common between both facilities include the lights and HVAC system. Unique end uses in this scenario include the coffee makers in Tony's Coffee Shop and the display coolers in Tina's Nursery.

Multiple Meters, Single Business

The third case we will look at is where you have a single business that is served by multiple meters as depicted in Figure 6-3. For this

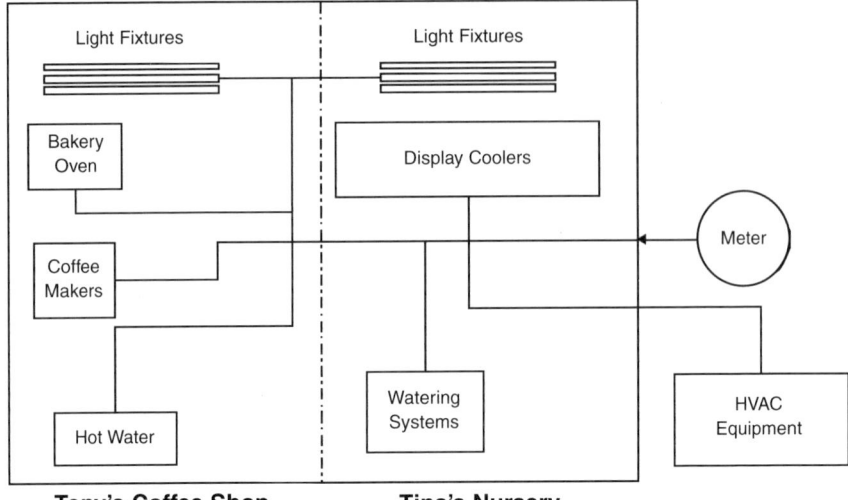

Figure 6-2. Single meter, multiple businesses.

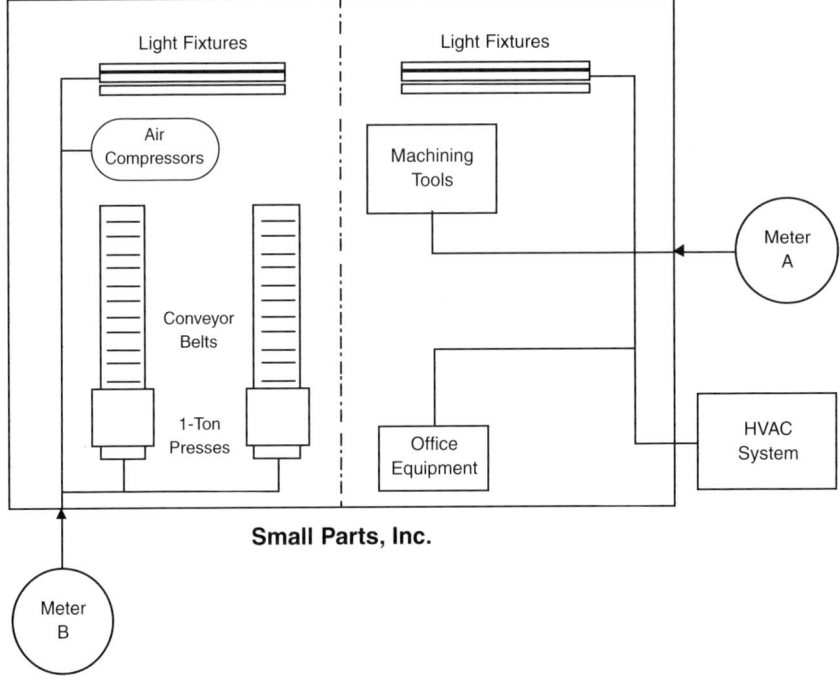

Figure 6-3. Multiple meters, single business.

case, we will use a company which produces parts for the automobile industry (Small Parts, Inc.). Over time, it has expanded a production line and added a second meter when the new services were installed. Once again, it's very important to distinguish which end uses are served by which energy meter as each meter may have belonged to a different sector or strata based upon the sample design for the study.

Multiple Meters, Multiple Businesses

The last case we will examine is for whole-building meters where you have multiple businesses that are served by multiple meters. For this discussion, we will use a strip mall that has a single meter for the air conditioning equipment, which serves all of the businesses in the mall. However, the remaining equipment is connected to a dedicated meter within each business as depicted in Figure 6-4.

As with the data collection in Chapter 5, when you analyze the data you will need to know which end uses are connected to each

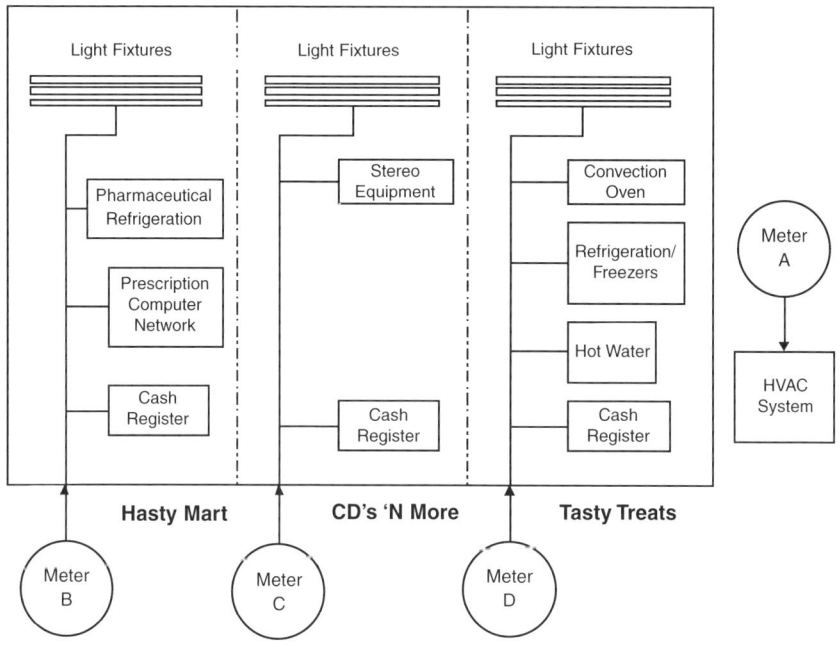

Figure 6-4. Multiple meters, multiple businesses.

whole-building meter that you are analyzing as well as which end uses are connected to the meter.

End-Use Metered Data

When you are trying to develop end-use load shapes it is no surprise that end-use metered data has a lot more "value" than whole-building load research data because, theoretically, the end-use metered data captures the true end-use consumption and patterns of use within a building. Additionally, if your sample design defined the premise as being a business it is generally easier to try to submeter those end uses that exist only within the business bounds. The biggest advantage of being able to isolate the business, even if it shared whole-building meters with other businesses, is that you will be able to use the population weights as represented by the sample design.

Unfortunately, even when dealing with end-use metered data you are not always guaranteed that you have data on all of the end uses within a facility or even that the end-use data you do have is a true representation of the end use which it supposedly was metered from. In this section, we present four scenarios that are encountered when analyzing end-use metered data as well as a discussion on how you can determine how "clean" your end-use metered data actually is.

Single Site, All End-Use Metered

From an analysis perspective, this type of end-use metered site is the most straightforward to work with. The first appealing thing about this scenario is that you have each end use individually metered in the building. Second, the population weight for this site should be the same as the sample design.

Note that in Figure 6-5, the submetering was implemented so each meter was able to record a unique end use. If you have a building which has dedicated circuits for each end use or you are willing to submeter individual end-use equipment, this is an achievable goal. Realize however, that you have to pay more (both in terms of budget and time) to meter individual equipment in a

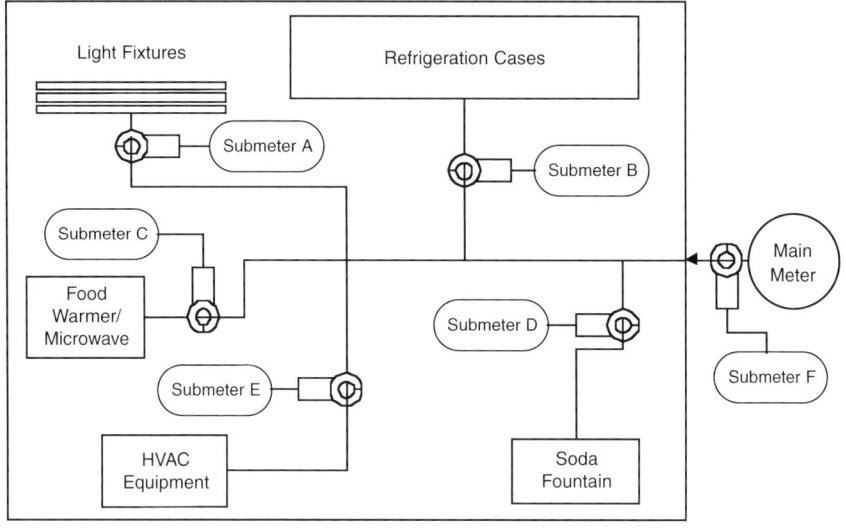

Figure 6-5. Single site, all end-use metered.

building rather than go in and simply submeter the individual circuits in the building.

Also note that we are using five submeters (A through E) to monitor the end-use equipment as well as a sixth meter (F) to monitor the whole-building load. The advantage of using submeter F is that it provides a quality control check on meters A through E as the sum of these meters should be equal to the monitored load at F.

Single Site, Some End-Uses Metered

In this case, we have a building which has only a portion of the end uses metered, perhaps due to budget constraints on the project.

In Figure 6-6 we once again examine the Hasty Mart which has some of its end uses metered. In this case, the only two end uses which are not being submetered are the soda fountain and the HVAC equipment. However, since you have metered all of the loads in the building at the primary meter, you can estimate the soda fountain and HVAC loads by subtracting all of the submetered loads from the load collected at the primary meter.

You could further examine this newly constructed end-use data and attempt to determine how much of this load belongs to the

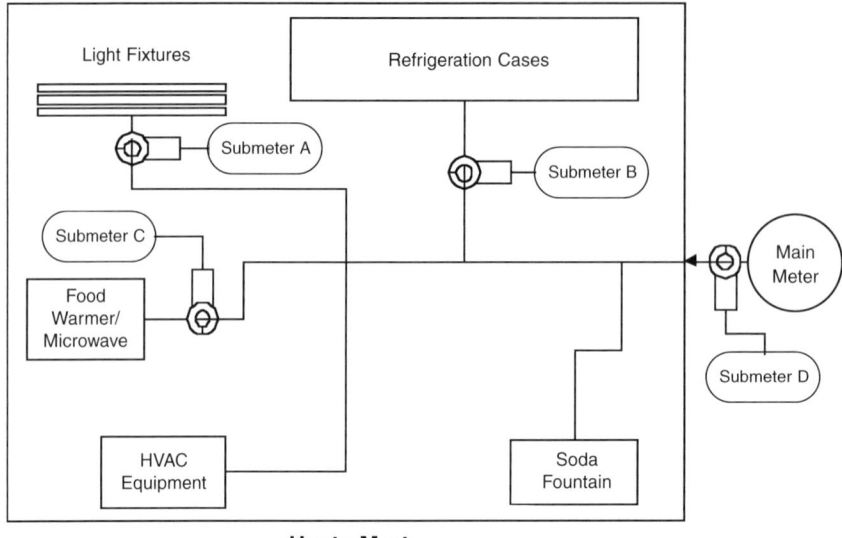

Figure 6-6. Single site, some end uses metered.

soda fountain and how much is actually the HVAC system. An alternative approach to separating these two end uses would be to simply assume that the soda fountain is negligible and to treat the entire load as the HVAC load.

Single Site, Some End-Uses Combined

In this third case, we will once again look at the Hasty Mart which has a different arrangement of electrical wiring as shown in Figure 6-7. On one electrical circuit we have the refrigeration equipment, the lighting fixtures, the food warmer, and microwave oven. On the second circuit we have the soda fountain and the HVAC equipment.

Once again, we may be able to assume that the soda fountain is an insignificant load and use the data collected at submeter B to approximate the HVAC end use. However, the data collected from submeter A is much more complicated.

Since each of the loads on the first circuit is a significant contributor to the load, we probably cannot simply assume them away. For circuit one, we will have to analyze an end-use load which is a combination of the refrigeration cases, lighting fixtures, and cooking appliances.

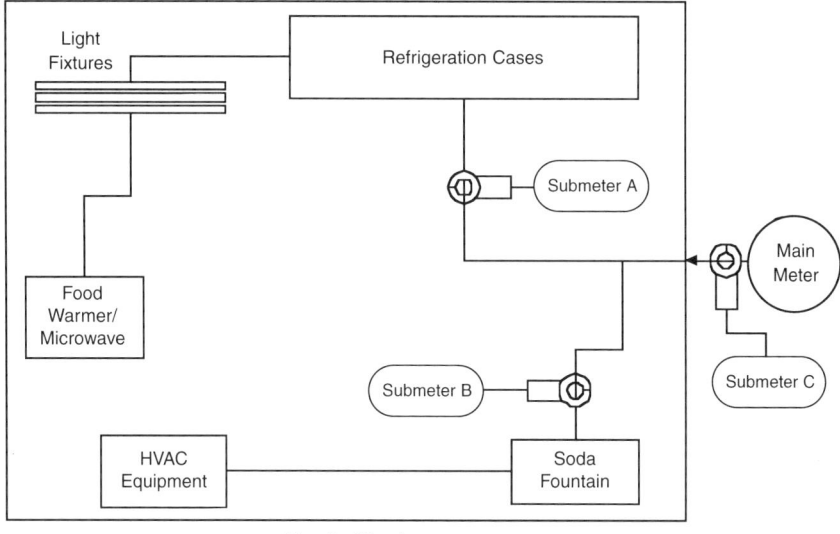

Figure 6-7. Single site, some end uses combined.

Single Site, Only Whole-Building Metered

In this final case, we appear to be dealing with a case which we covered earlier as the "single meter, single business" case. However, if there is only one end use connected to the meter we would actually be dealing with the "single site, some end uses metered" case. However, as this will rarely occur we will generally treat this site as only having whole-building load research data, and we need to know which end uses are connected to the meter, such as is presented in Figure 6-8.

How Clean Is Clean?

In the previous sections of this chapter, we discussed whether some of the end-use data is clean and whether individual loads are significant or insignificant; but what exactly was meant by these various terminologies?

When you are utilizing end-use metered data you commonly have data which is not a pure end use. Expressed another way, you may have data from submetering in a facility that is supposed to

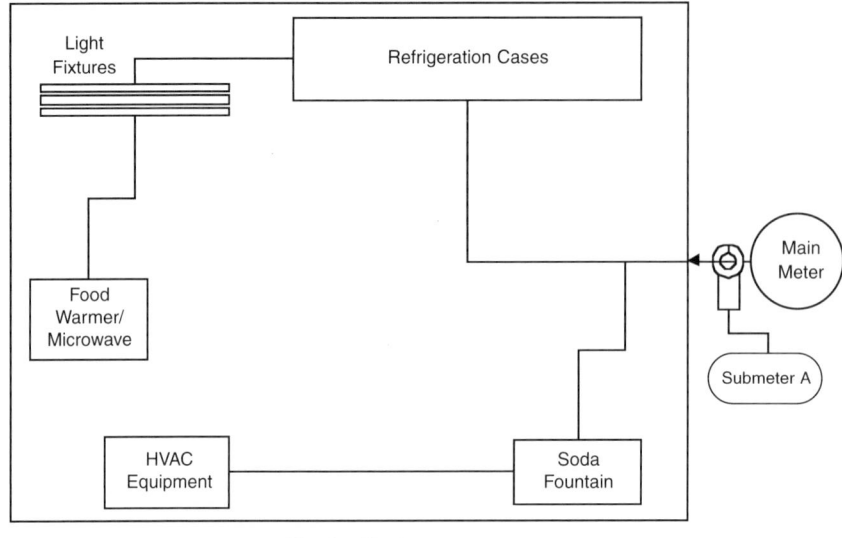

Figure 6-8. Single site, only whole-building metered.

represent a single end use in which equipment from more than one end use is represented.

Generally, you need to decide how much contamination you will allow in your end-use metered data. Some metering studies I've encountered have defined clean data as data in which 95 percent or more of the load is contributed by a single end use. Furthermore, we consider a load significant if its maximum peak load is large enough to cause the data *not* to be considered clean. The cleanness of a circuit can be determined using Equation 6-1 as shown below:

$$Clean\ Fraction = \frac{\sum Loads_{end\ use}}{\sum Loads} \quad \textbf{(Equation 6-1)}$$

where:

Clean Fraction = fractional percent of metered data that belongs to the end use

$Loads_{end\ use}$ = loads which belong to the desired end use class

Loads = all loads on the submeter.

Table 6-1. Average loads for equipment being submetered.

Equipment	End-Use Category	Average Load (kW)
Lighting fixtures	Lighting	32.5
Display cases	Lighting	2.3
Cash registers	Miscellaneous	6.7
Plug loads	Miscellaneous	1.4
Total	N/A	42.9

Example 6-1: Cleanness of Submetered Data

You have a retail store which has submetered data for one of the circuits in the store. During an average day, the following loads are observed to occur in the circuit for the equipment listed in Table 6-1.

Determine how clean this circuit is and whether each piece of equipment is significant or insignificant assuming a predefined goal that circuits need to be 90 percent pure or clean. Additionally, the metering plan informs you that this data is intended to represent the lighting end-use category.

Using Equation 6-1 we can determine that the clean fraction of the submeter which is contributed by the lighting end use is:

$$Clean\ Fraction = \frac{32.5 + 2.3}{42.9} = 0.81$$

This implies that this circuit fails the desired goal of being 90 percent clean and therefore must be analyzed as a combination of end uses—the lighting and miscellaneous categories—or that some other method must be implemented to separate the miscellaneous loads from the lighting loads in this data.

Since we know that the criteria for cleanness is the 90 percent level, it follows that significant loads are any combination of loads which contribute at least 10 percent of the total load for the submeter. For this particular submetering example, a load may be considered significant if it accounts for 10 percent of the total load or 4.3 kW.

Based on this result equipment that represent significant loads includes the lighting fixtures and the cash registers; while the display cases and the plug loads represent insignificant loads. Realize however that you don't want to drop the display case load because it belongs to the lighting end-use category. Furthermore, while you could drop the plug loads from this analysis you would still be left with the cash register loads which belong to the miscellaneous end use, which means you would still have an end use which was not clean.

Condensing the Data

Depending upon which type of metering you use and the time increment at which you are collecting end-use metered data, you could end up with a tremendous amount of data. In electric utilities it is typical to collect and store whole-building metered data in 15-minute increments for every hour of the year, which amounts to more than 35,000 data points for each building. If you were also submetering end-use data for nine end uses for the same yearly period, you would have more than 300,000 data points to manage. Clearly, you probably don't want to have to deal with all of this data during the analysis stage. As discussed in Chapter 2, one method of dealing with all of this data is to condense it into daytypes of interest. Chapter 2 presents common daytype definitions for the 16, 36, and 48 daytype schemes.

Example 6-2: Developing Daytype Data

Condense data from six separate days into an average weekday daytype for the summer season. We will assume that the summer season includes the months of April through September. Hourly data for the six daytypes are listed in Table 6-2. To obtain the average loads, we calculate the arithmetic mean of the loads from all six days for each hour. The results for the average weekday are presented as Figure 6-9.

If you wanted to select the average peak summer weekday, you would choose the data from July 16, as this has the largest load among all of the summer daytypes. If you have a larger number of days in your analysis you may wish to calculate the peak day as the arithmetic mean of the top 10 percent of the days that have the highest load.

Table 6-2. Hourly load for five summer days.

Hour	April 16	May 24	June 12	July 16	Aug. 8	Sept. 9
1	2.32	5.46	1.50	4.13	0.21	2.09
2	3.45	3.30	2.92	2.43	1.58	2.66
3	5.82	4.21	5.19	1.74	2.43	6.04
4	5.07	4.78	8.44	2.58	5.45	5.06
5	4.29	6.05	8.37	4.17	4.78	4.94
6	6.11	5.25	6.44	7.31	5.33	11.84
7	11.89	13.72	7.66	12.60	6.78	16.94
8	20.55	19.44	7.09	20.73	12.36	18.13
9	18.27	19.47	12.18	22.00	23.10	17.88
10	18.02	16.50	15.00	28.00	21.14	18.62
11	11.55	15.30	23.22	38.16	21.18	28.29
12	10.99	14.20	28.25	38.00	19.30	27.00
13	11.20	17.60	31.41	38.67	31.17	25.40
14	13.49	13.43	32.02	37.32	38.45	14.06
15	14.12	17.27	31.07	29.83	35.88	12.75
16	16.65	19.39	28.14	31.40	30.23	7.74
17	15.79	23.95	25.44	28.18	17.04	11.00
18	14.19	19.59	15.52	21.52	10.24	12.62
19	10.39	15.13	8.30	14.77	3.65	9.88
20	2.71	8.92	2.02	8.71	6.17	7.88
21	3.15	8.44	2.35	6.74	5.27	6.56
22	3.62	5.18	1.15	4.66	3.69	2.75
23	4.11	4.07	0.75	1.47	1.21	1.12
24	2.70	5.37	0.21	0.26	0.62	0.72

Aggregating the Data

Once we have all of the whole-building or end-use metered data, we need to analyze them in some fashion. There are three ways that we can combine these data from individual buildings to derive information about the utility population as a whole: (1) aggregate all of data to the population level, (2) produce buildings which are typical of the population, and (3) simply average the whole-building or end-use metered data to develop average load shapes.

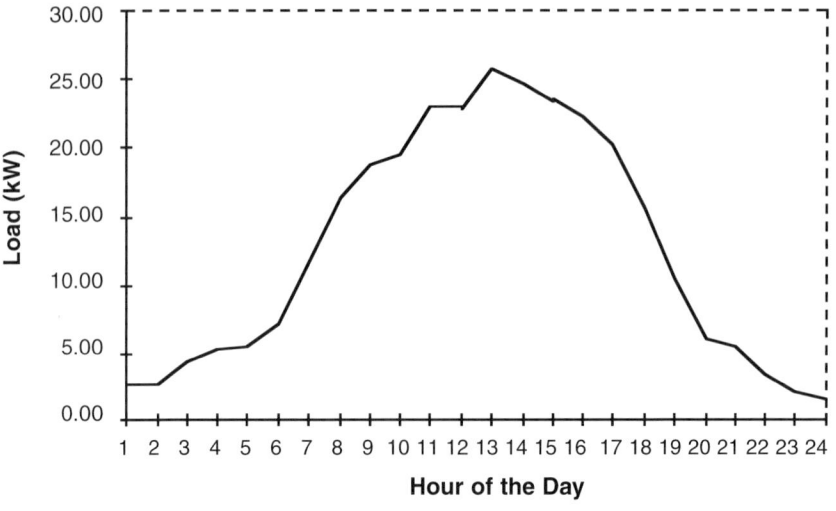

Figure 6-9. Average summer weekday example.

Aggregate End-Use Data to Population Level

If you are doing a study to predict the load shapes for your population or a segment of your population this aggregation approach is most applicable. The basic method for this approach is to multiply the end-use or whole-building load shapes you obtain for a single building by the population weight for that particular building, as shown in Equation 6-2 to obtain the population representation of that building.

$$Load_{Pop-i,eu} = \sum Load_{i,eu} \bullet Weight_i \qquad \textbf{(Equation 6-2)}$$

where:

$Load_{Pop-i,eu}$ is the population representation of the load for building i and end-use eu

$Load_{i,eu}$ is the load for building i and end-use eu

$Weight_i$ is the population weight for building i.

Repeat this process for all of the buildings and end uses in your study and you should have a representation of all your sectors (or building types) at the population level as shown in Equation 6-3.

$$Load_{Pop,eu} = \sum_i Load_{i,eu} \cdot Weight_i \quad \textbf{(Equation 6-3)}$$

where:

$Load_{Pop,eu}$ is the population representation of end-use eu

$Load_{i,eu}$ is the load for building i and end-use eu

$Weight_i$ is the population weight for building i.

Example 6-3: Aggregating End-Use Data

Aggregate a set of end-use metered data to the grocery population level. We have list a set of typical end-use metered data for five sites that represent the grocery sector in Table 6-3. Note that in Table 6-3, we have listed a site identification number, the sample design stratum of the site, and the hourly loads for four simplified end uses which we previously condensed into an average summer weekday daytype.

In order to aggregate the results to the grocery population level, we first need to know the population weights for each stratum. According to our sample design the population weights are 24 and 9, respectively. These imply that each grocery in stratum 1 represents 24 grocery stores in the population and each grocery store in stratum 2 represents 9 groceries.

Applying Equation 6-3 to the building data, we generate the results for the grocery sector as presented in Table 6-4.

Aggregating Whole-Building Load Research Data

For the case where you only have whole-building load research data and no end-use metered data, Equation 6-3 reduces to the form shown in Equation 6-4.

$$Load_{Pop} = \sum_i Load \cdot Weight_i \quad \textbf{(Equation 6-4)}$$

where:

$Load_{Pop}$ is the population representation of the whole-building load for building i

$Load_i$ is the whole-building load for building i

$Weight_i$ is the population weight for building i.

Table 6-3. Average summer weekday end-use loads.

			Loads by Hour of the Day											
Site ID	Stratum	End Use	1	2	3	4	5	6	7	8	9	10	11	12
3409	1	Lights	3.7	3.6	3.9	4.0	3.7	3.9	8.0	12.0	11.6	11.0	10.8	11.8
3409	1	Refrigeration	13.2	3.7	14.8	14.2	4.0	6.7	8.1	22.2	12.0	11.2	22.2	19.1
3409	1	Miscellaneous	2.8	4.6	2.5	4.6	2.3	2.3	6.7	2.1	8.9	2.5	2.1	2.2
3409	1	HVAC	0.0	0.8	0.1	0.2	0.6	0.4	1.7	0.9	6.4	4.7	1.3	3.7
4607	1	Lights	14.9	13.8	13.5	13.6	13.6	15.0	13.8	13.1	14.6	14.0	13.4	14.9
4607	1	Refrigeration	19.1	16.3	18.5	19.6	17.9	18.8	15.6	15.0	21.1	21.4	23.8	22.6
4607	1	Miscellaneous	6.5	10.4	12.8	2.5	5.7	7.1	2.7	2.4	9.6	2.3	6.0	9.7
4607	1	HVAC	1.3	1.9	2.2	0.9	1.5	1.6	1.4	2.3	6.0	3.3	5.3	10.8
5211	1	Lights	8.3	8.2	8.7	11.1	11.9	11.0	17.9	17.4	16.2	17.2	17.0	17.4
5211	1	Refrigeration	17.4	16.2	18.2	12.2	17.2	12.9	19.6	19.9	26.4	19.5	23.7	22.5
5211	1	Miscellaneous	4.4	2.0	3.4	5.1	4.2	5.6	4.8	3.1	8.5	3.7	9.8	10.0
5211	1	HVAC	0.4	0.3	0.4	1.4	1.1	1.3	2.1	3.1	5.0	6.6	10.6	13.0
6426	2	Lights	27.5	28.3	29.7	27.4	29.6	29.4	27.2	28.9	27.3	27.4	27.5	28.9
6426	2	Refrigeration	35.9	28.4	40.4	38.6	36.1	39.0	37.1	31.5	34.1	35.0	35.3	37.3
6426	2	Miscellaneous	16.7	7.2	4.9	16.3	23.5	9.9	4.4	16.3	13.7	7.3	6.0	8.2
6426	2	HVAC	2.9	2.6	1.9	3.4	5.3	2.6	2.1	8.5	10.5	10.1	11.2	14.9
7899	2	Lights	4.2	4.3	4.2	4.1	67.7	68.1	64.7	65.4	64.1	69.5	69.9	66.3
7899	2	Refrigeration	5.7	5.8	15.5	4.2	77.0	72.5	68.3	76.8	65.4	78.6	73.3	71.8
7899	2	Miscellaneous	3.3	2.3	4.0	5.7	7.8	6.2	9.3	3.2	13.9	14.6	31.6	11.5
7899	2	HVAC	0.5	0.4	0.0	1.1	5.6	4.9	6.4	8.8	19.9	26.4	46.1	34.0

Metered Data Analysis

Loads by Hour of the Day

Site ID	Stratum	End Use	13	14	15	16	17	18	19	20	21	22	23	24
3409	1	Lights	11.6	11.5	11.3	11.1	11.5	11.7	11.3	11.7	11.0	3.7	3.9	3.8
3409	1	Refrigeration	21.6	16.9	18.8	15.7	14.0	19.8	22.0	19.2	20.8	15.5	15.0	3.9
3409	1	Miscellaneous	5.3	2.1	2.7	8.4	5.0	2.6	8.3	2.5	3.1	2.1	4.8	3.7
3409	1	HVAC	5.3	4.6	4.0	9.3	6.6	2.4	3.4	1.3	0.8	0.4	0.2	0.7
4607	1	Lights	14.1	14.2	13.1	14.9	14.9	14.2	14.9	14.2	13.8	14.9	13.8	14.2
4607	1	Refrigeration	25.5	19.0	20.2	26.5	24.0	18.2	24.2	19.7	20.8	21.3	16.8	21.5
4607	1	Miscellaneous	3.1	3.1	4.6	3.1	6.4	3.6	9.6	6.9	2.1	5.3	11.4	4.5
4607	1	HVAC	3.9	6.8	6.7	3.8	6.4	4.8	5.0	3.1	1.2	1.7	2.5	1.0
5211	1	Lights	16.5	17.9	16.4	16.6	17.7	16.5	16.1	16.4	11.3	11.7	11.4	8.5
5211	1	Refrigeration	18.5	26.9	16.4	20.6	18.3	17.2	16.4	18.6	13.1	18.2	15.5	16.2
5211	1	Miscellaneous	2.9	11.3	6.8	5.0	11.5	10.3	2.4	4.1	3.1	7.3	4.3	4.8
5211	1	HVAC	9.0	14.0	13.1	9.0	13.8	10.0	4.1	3.2	1.7	1.8	1.2	0.7
6426	2	Lights	28.7	27.9	28.0	28.1	28.6	29.4	29.1	28.1	29.7	28.2	27.1	27.3
6426	2	Refrigeration	37.4	31.9	30.5	33.9	28.7	31.7	34.2	38.1	39.0	39.8	32.8	28.3
6426	2	Miscellaneous	12.6	3.9	22.7	16.4	17.6	7.1	8.2	4.4	13.0	3.8	3.4	18.8
6426	2	HVAC	19.9	14.0	30.8	22.1	22.0	11.4	8.1	3.8	5.1	2.2	2.1	4.2
7899	2	Lights	69.1	69.7	67.8	66.9	69.8	67.8	69.0	67.0	65.2	63.5	4.3	4.4
7899	2	Refrigeration	69.2	70.2	67.9	78.5	70.2	73.3	72.0	77.2	72.0	74.5	6.3	9.6
7899	2	Miscellaneous	5.4	26.2	14.3	12.7	6.0	19.4	33.8	5.4	51.3	44.7	5.1	2.1
7899	2	HVAC	35.1	54.1	42.0	32.3	28.1	27.8	26.7	9.5	17.7	12.8	0.9	0.2

Table 6-4. Grocery sector end-use loads (kW).

End Use	Loads by Hour of the Day											
	1	2	3	4	5	6	7	8	9	10	11	12
Lights	930.9	907.8	931.5	972.3	1576.5	1595.1	1779.9	1868.7	1840.2	1884.9	1865.4	1915.2
Refrigeration	1567.2	1176.6	1739.1	1489.2	1956.3	1925.1	1987.8	2345.1	2323.5	2272.8	2650.2	2522.7
Miscellaneous	508.8	493.5	528.9	490.8	574.5	504.9	464.1	357.9	896.4	401.1	768.0	702.9
HVAC	71.4	99.0	81.9	100.5	174.9	146.7	201.3	306.9	691.2	678.9	928.5	1100.1
	13	14	15	16	17	18	19	20	21	22	23	24
Lights	1893.0	1924.8	1841.4	1877.4	1944.0	1892.4	1898.1	1871.1	1720.5	1552.5	981.0	921.3
Refrigeration	2533.8	2426.1	2215.2	2518.8	2241.3	2269.8	2458.2	2417.7	2311.8	2348.7	1487.1	1339.5
Miscellaneous	433.2	666.9	671.4	657.9	762.0	634.5	865.2	412.2	777.9	789.3	568.5	500.1
HVAC	931.8	1222.5	1226.4	1020.0	1094.1	765.6	613.2	302.1	294.0	228.6	120.6	97.2

Example 6-4: Aggregating Load Research Data

For the buildings previously presented in Example 6-4, calculate the total load for the grocery sector. The total loads for each of the buildings is shown in Table 6-5. Using the data, we apply the population and total loads for each hour and building to Equation 6-4 to obtain the population representation of the total load shown in Table 6-5.

Aggregate End-Use Data to Typical Building

If you don't want to estimate population totals, but rather want to estimate the load shapes for a typical customer in a segment of your population, this aggregation approach is most applicable. The basic method for this approach is to sum the multiple of the end-use or whole-building load shapes you obtain for a single building by the population weight for that particular building, and divide the total by the sum of the population weights for each strata as shown in Equation 6-5 to obtain the population representation of that building.

$$Load_{Typ-i,eu} = \frac{\sum_i Load_{i,eu} \cdot Weight_i}{\sum_i Weight_i}$$ (Equation 6-5)

where:

$Load_{Typ-i,eu}$ is the typical building representation of the load for building i and end-use eu

$Load_{i,eu}$ is the load for building i and end-use eu

$Weight_i$ is the population weight for building i.

Example 6-5: Developing Typical End-Use Load Shapes

Now, we will produce typical end-use load shapes for the buildings presented in Example 6-3. Using the data contained in Table 6-3 and applying Equation 6-5, we produce the grocery sector typical load shapes which are shown in Table 6-6. It's important to understand that typical load shapes are not the same thing as average load shapes, which we present later in this chapter.

Table 6-5. Total loads for sites in grocery sector.

			Loads by Hour of the Day										
Site ID	Stratum	Weight	1	2	3	4	5	6	7	8	9	10	11	12
3409	1	24	19.7	12.7	21.3	23.0	10.6	13.3	24.5	37.2	38.9	29.4	36.4	36.8
4607	1	24	41.8	42.4	47.0	36.6	38.7	42.5	33.5	32.8	51.3	41.0	48.5	58.0
5211	1	24	30.5	26.7	30.7	29.8	34.4	30.8	44.4	43.5	56.1	47.0	61.1	62.9
6426	2	9	83.0	66.5	76.9	85.7	94.5	80.9	70.8	85.2	85.6	79.8	80.0	89.3
7899	2	9	13.7	12.8	23.7	15.1	158.1	151.7	148.7	154.2	163.3	189.1	220.9	183.6

Site ID	Stratum	Weight	13	14	15	16	17	18	19	20	21	22	23	24
3409	1	24	43.8	35.1	36.8	44.5	37.1	36.5	45.0	34.7	35.7	21.7	23.9	12.1
4607	1	24	46.6	43.1	44.6	48.3	51.7	40.8	53.7	43.9	37.9	43.2	44.5	41.2
5211	1	24	46.9	70.1	52.7	51.2	61.3	54.0	39.0	42.3	29.2	39.0	32.4	30.2
6426	2	9	98.6	77.7	112.0	100.5	96.9	79.6	79.6	74.4	86.8	74.0	65.4	78.6
7899	2	9	178.8	220.2	192.0	190.4	174.1	188.3	201.5	159.1	206.2	195.5	16.6	16.3

(cont'd)

Table 6-5. Estimated total load for grocery sector (cont'd).

	Loads by Hour of the Day											
	1	2	3	4	5	6	7	8	9	10	11	12
Load (kW)	3078.3	2676.9	3281.4	3052.8	4282.2	4171.8	4433.1	4878.6	5751.3	5237.7	6212.1	6240.9
	13	14	15	16	17	18	19	20	21	22	23	24
Load (kW)	5791.8	6240.3	5954.4	6074.1	6041.4	5562.3	5834.7	5003.1	5104.2	4919.1	3157.2	2858.1

Table 6-6. Grocery sector typical load end-use load shapes.

End Use	Loads by Hour of the Day											
	1	2	3	4	5	6	7	8	9	10	11	12
Lights	10.3	10.1	10.4	10.8	17.5	17.7	19.8	20.8	20.4	20.9	20.7	21.3
Refrigeration	17.4	13.1	19.3	16.5	21.7	21.4	22.1	26.1	25.8	25.3	29.4	28.0
Miscellaneous	5.7	5.5	5.9	5.5	6.4	5.6	5.2	4.0	10.0	4.5	8.5	7.8
HVAC	0.8	1.1	0.9	1.1	1.9	1.6	2.2	3.4	7.7	7.5	10.3	12.2

End Use	13	14	15	16	17	18	19	20	21	22	23	24
Lights	21.0	21.4	20.5	20.9	21.6	21.0	21.1	20.8	19.1	17.3	10.9	10.2
Refrigeration	28.2	27.0	24.6	28.0	24.9	25.2	27.3	26.9	25.7	26.1	16.5	14.9
Miscellaneous	4.8	7.4	7.5	7.3	8.5	7.1	9.6	4.6	8.6	8.8	6.3	5.6
HVAC	10.4	13.6	13.6	11.3	12.2	8.5	6.8	3.4	3.3	2.5	1.3	1.1

Aggregate Whole-Building Data to Typical Building

The equation to perform this operation is a modification of Equation 6-5, where instead of applying it to an end use, the equation is applied to the whole-building load as shown in Equation 6-6.

$$Load_{Typ-i,bldg} = \frac{\sum_i Load_{i,bldg} \cdot Weight_i}{\sum_i Weight_i} \quad \textbf{(Equation 6-6)}$$

where:

$Load_{Typ-i,bldg}$ is the typical building representation of the total load for building i

$Load_{i,bldg}$ is the whole-building load for building i

$Weight_i$ is the population weight for building i.

Example 6-6: Developing Typical Whole Building Load Shapes

Calculate the typical whole-building load for the grocery sector using the data presented in Table 6-5. After we apply Equation 6-6 to the data we obtained the results shown in Table 6-7.

Aggregate End-Use Metered Data to Average Load Shape

The primary difference between developing typical and average load shapes is that typical load shapes include population weights in the calculation. To calculate average load shapes, however, we can simply calculate the arithmetic mean of a group of end uses to find the average end use as shown in Equation 6-7.

$$Load_{Avg-i,eu} = \frac{\sum_i Load_{i,eu} \cdot Weight_i}{n} \quad \textbf{(Equation 6-7)}$$

where:

$Load_{Avg-i,eu}$ is the typical building representation of the total load for building i

$Load_{i,eu}$ is the whole-building load for building i

$Weight_i$ is the population weight for building i

n is the number of buildings in your sample.

Table 6-7. Typical whole-building load for grocery sector.

	Loads by Hour of the Day											
	1	2	3	4	5	6	7	8	9	10	11	12
Load (kW)	34.2	29.7	36.5	33.9	47.6	46.4	49.3	54.2	63.9	58.2	69.0	69.3
	13	14	15	16	17	18	19	20	21	22	23	24
Load (kW)	64.4	69.3	66.2	67.5	67.1	61.8	64.8	55.6	56.7	54.7	35.1	31.8

Example 6-7: Developing Average End-Use Load Shapes

Calculate the average end-use loads for the grocery sector using the data presented in Table 6-4. After we apply Equation 6-7 to the data we obtained the results shown in Table 6-8.

Aggregate Whole-Building Data to Average Load Shape

As with average end-use load shapes, the difference between developing typical and average load shapes is that the typical load shapes include the population weights in the calculation. To calculate average whole-building loads, however, we can simply calculate the arithmetic mean of a group of whole-building loads to find the average whole-building load as shown in Equation 6-8.

$$Load_{Avg-i,bldg} = \frac{\sum_i Load_{bldg} \cdot Weight_i}{n} \quad \textbf{(Equation 6-8)}$$

where:

$Load_{Avg-i,bldg}$ is the typical building representation of the total load for building i

$Load_{i,bldg}$ is the whole-building load for building i

$Weight_i$ is the population weight for building i

n is the number of buildings in your sample.

Example 6-8: Developing Average Whole-Building Load Shapes

Calculate the average whole-building loads for the grocery sector using the data presented in Table 6-5. After we apply Equation 6-8 to the data we obtained the results shown in Table 6-9.

In the next three chapters, we will present additional methods for calculating and manipulating end-use load shapes, including: engineering analysis, statistical analysis, and hybrid statistical/engineering models.

Table 6-8. Average end-use load shapes.

Loads by Hour of the Day

End Use	1	2	3	4	5	6	7	8	9	10	11	12
Lights	11.7	11.6	12.0	12.0	25.3	25.5	26.3	27.4	26.8	27.8	27.7	27.9
Refrigeration	18.3	14.1	21.5	17.8	30.4	30.0	29.7	33.1	31.8	33.1	35.7	34.7
Miscellaneous	6.7	5.3	5.5	6.8	8.7	6.2	5.6	5.4	10.9	6.1	11.1	8.3
HVAC	1.0	1.2	0.9	1.4	2.8	2.2	2.7	4.7	9.6	10.2	14.9	15.3

End Use	13	14	15	16	17	18	19	20	21	22	23	24
Lights	28.0	28.2	27.3	27.5	28.5	27.9	28.1	27.5	26.2	24.4	12.1	11.6
Refrigeration	34.4	33.0	30.8	35.0	31.0	32.0	33.8	34.6	33.1	33.9	17.3	15.9
Miscellaneous	5.9	9.3	10.2	9.1	9.3	8.6	12.5	4.7	14.5	12.6	5.8	6.8
HVAC	14.6	18.7	19.3	15.3	15.4	11.3	9.5	4.2	5.3	3.8	1.4	1.4

Table 6-9. Average whole-building load shapes.

<table>
<tr><th colspan="13">Loads by Hour of the Day</th></tr>
<tr><th></th><th>1</th><th>2</th><th>3</th><th>4</th><th>5</th><th>6</th><th>7</th><th>8</th><th>9</th><th>10</th><th>11</th><th>12</th></tr>
<tr><td>Load (kW)</td><td>37.7</td><td>32.2</td><td>39.9</td><td>38.0</td><td>67.3</td><td>63.8</td><td>64.4</td><td>70.6</td><td>79.0</td><td>77.3</td><td>89.4</td><td>86.1</td></tr>
<tr><th></th><th>13</th><th>14</th><th>15</th><th>16</th><th>17</th><th>18</th><th>19</th><th>20</th><th>21</th><th>22</th><th>23</th><th>24</th></tr>
<tr><td>Load (kW)</td><td>82.9</td><td>89.2</td><td>87.6</td><td>87.0</td><td>84.2</td><td>79.8</td><td>83.8</td><td>70.9</td><td>79.2</td><td>74.7</td><td>36.6</td><td>35.7</td></tr>
</table>

7
Engineering Analysis

The House That Jack Built

The topic of engineering analysis is quite broad and can cover everything from simple engineering equations and algorithms to more complex building simulation models. For the purposes of this book I would like to differentiate between building simulation models and load shape analysis tools.

Building Simulation Models

Building simulation models were originally developed to size the heating and cooling requirements for commercial buildings in various climates and using a variety of heating and cooling equipment. One of the early models developed at Lawrence Berkeley Laboratory (LBL) which is still used extensively was DOE-2 (since it was funded by the U.S. Department of Energy).

The World Wide Web page at LBL describes DOE-2 as "a public-domain computer program that performs an hour-by-hour simulation of a building's expected energy use and energy cost given a description of the building's climate, architecture, materials, operating schedules, and HVAC equipment. DOE-2 is widely used in the United States and 42 other countries to design energy-efficient buildings, to analyze the impact of new technologies, and to develop energy conservation standards." These tools have historically been very complicated to use and difficult to understand.

Developers at LBL are currently working on POWERDOE, which will be a Windows-based software program that will run on the PC and will supposedly offer a much easier to use user-interface as well as other features. Another highly sophisticated building energy model in use today is BLAST, which was funded by the Construction Engineering Research Laboratory. These models both use hourly weather data including outdoor temperature, solar radiation, and other parameters.

Another organization which has historically been very active in developing building simulation methods and computer models is the American Society of Heating, Refrigerating and Air Conditioning Engineers, Inc. (ASHRAE). In 1967, ASHRAE introduced the Total Equivalent Temperature Difference/Time-Averaging method (TETD/TA). This method provides simple equations that can be used to calculate the solar radiation, conduction heat gain, and internal heat gains. In the mid 1970s, ASHRAE published the Cooling Load Temperature Difference/Solar Cooling Load/ Cooling Load Factor method (CLTD/SCL/CLF) to determine cooling loads for a building. Both the TETD/TA and CLTD/SCL/CLF methods were designed so that HVAC system designers could perform manual calculations to determine what size air-conditioning equipment was required for a given building. In the early 1980s, ASHRAE published the Simplified Bin Energy Analysis method (commonly known as the Bin method) which uses annual weather data, collapsed into temperature bins, to determine loads. Several computer models were developed and widely used that used these simplified methods to provide a relatively quick answer to how much heating and cooling load a particular building had. One of these models was the ASHRAE Simplified Energy Analysis Method. In the late 1980s, ASHRAE published their Transfer Function Method (TFM) of determining cooling and heating loads in a building. TFM is a complex method designed for being implemented into computer programs (McQuiston and Spitler, 1992).

While building simulation models tend to be complex and somewhat difficult to use, they share a primary common goal of trying to determine the peak heating and cooling loads for a given set of building parameters. Unfortunately, these models quite often fall short when trying to perform load shape analysis on buildings.

Load Shape Analysis Models

Since building simulation models have an emphasis on the heating and cooling loads in a building, they quite often oversimplify or even ignore some of the other loads in a building. On the other hand, based upon my experience, load shape analysis tools tend to be more sophisticated in determining the underlying loads in a building and less concerned about the heating and cooling loads in a building. One of the reasons for this is that most load shape analysis tends to be more exploratory in nature where building simulation models have a single goal of determining the peak heating and cooling loads. If you are trying to use a building simulation model, your inputs and outputs may be constrained which limits your ability to analyze the data. In determining which approach you should take on a specific project you should analyze the unique needs of that project.

End-Use Models

This section discusses what variables are necessary to model various end uses and some of the simple end-use models that can be used for analysis. This section will describe models for the following end uses:

- indoor lighting
- exterior lighting
- miscellaneous equipment
- domestic hot water
- commercial cooking
- refrigeration/freezers
- ventilation
- space heating
- air conditioning

Indoor Lighting

Indoor lighting includes light fixtures in the ceiling, floor lamps, desk lamps, and any other lighting source used inside a building. There are several factors which affect how much of the lighting is on in a building or room at any given time, including what type of lighting controls are used and whether or not the room is occupied at the time. Before we get into the different types of models which may be used to simulate lighting load it should be pointed out that there are two broad approaches for determining the power consumption of indoor lights in a facility: (1) using a wattage density or (2) individually counting every lighting fixture and recording its specific rated wattage or actual wattage based on laboratory test results. My preference is to use the fixture counting method when time allows. However, if you are trying to develop load shapes for buildings which are under construction then you would probably use the proposed wattage density for the building in your load shape calculations.

As with most end-use models, lighting models can range in complexity from a simple "box" model to more complicated models which incorporate occupancy sensors, daylight sensors, and other factors.

Box Model

In the basic box model we assume that all of the lights in a facility start and stop at a fixed time every business day. Note that the times do not have to be the same for every day of the week but are consistent from week to week. The lights are assumed to be on from the start time until the stop time and off during other periods of the day.

Lighting Example 7-1: Box Model

Assume that we have a commercial laundry which is open from 6:00 A.M. to 6:00 P.M. Monday through Friday and from 8:00 A.M. to 5:00 P.M. on Saturday, and that the connected load of the lights in the facility is 3.4 kW. The lighting loads for the weekdays and Saturday are both presented as Figure 7-1.

Space Model

The lighting space model is an extension of the box model. Where the box model assumes a fixed schedule for the entire building, the

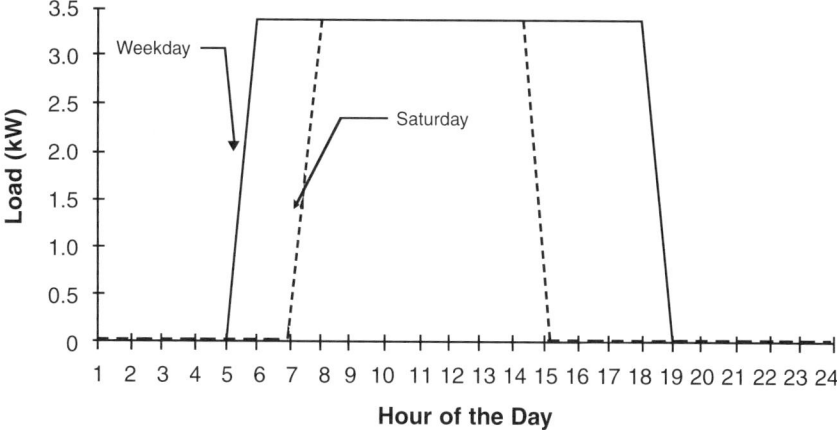

Figure 7-1. Box model lighting loads.

space model allows individual schedules to be assigned to various spaces within the facility. Whether or not you use the space model over the box model is merely a function of how the lights in a facility are actually controlled. If all of the lights in a space were connected to a single light switch or if the first person to arrive in the morning turned on all of the lights and the last person to leave in the evening turned all of the lights off, both of which may occur in a small establishment, then the box model is appropriate. However, if different spaces generally have lights on at unique times then the space model is most appropriate.

Lighting Example 7-2: Space Model

We once again will use the commercial laundry, but we will now break it down by defining unique spaces within the building which have different operating schedules. The cashier area has a connected lighting load of 0.5 kW and is open from 6:00 A.M. to 6:00 P.M. Monday through Friday; the offices in the building also have a connected load of 0.5 kW and are utilized from 1:00 P.M. to 4:00 P.M. and the area of the building where the laundry is actually cleaned is open from 6:00 A.M. to 1:00 P.M. and has a connected load of 2.4 kW. The lighting loads for weekdays is presented as Figure 7-2.

Occupancy Sensor Model

Besides the manual control of lighting that occurs in facilities, there are also automatic controls which change the lighting load in a

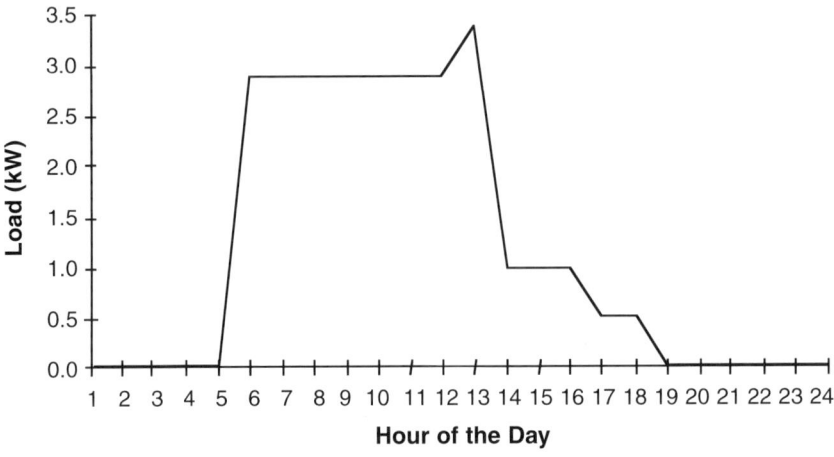

Figure 7-2. Space model lighting loads.

facility. The occupancy sensor model can provide a simple means of determining whether a light in a given space is on or off at a given time. The rationale of the basic occupancy sensor model is that the lights in a space are on if it's occupied and off at all other times.

The trickiest part is determining how often spaces such as conference rooms and even private offices are occupied on average. Additionally, within any company the occupancy patterns in each office will vary depending upon the patterns or duties of the occupant, therefore the job of the person collecting data about a site is to determine how often each space type which has occupancy sensors is occupied.

If you are writing a load shape development tool, the simplest way to implement the occupancy sensor is to multiply the connected load by the occupancy indicator for the desired hour as shown in Equation 7-1.

$$Load_i = Load_{connect} \cdot Occupancy\ Indicator_i \quad \textbf{(Equation 7-1)}$$

where:

$Load_i$ = actual lighting load at hour i

$Load_{connect}$ = connected lighting load for the space

$Occupancy\ Indicator_i$ = fraction of hour that the space is occupied.

Equation 7-1 can be applied to both single-person offices and spaces which have multiple persons such as shared offices, conference rooms, and cafeterias.

Occupancy Sensor Example
Say you have an office building which has nominal operating hours of 8:00 A.M. to 5:00 P.M. and five employees. If each employee has his/her own office, with an occupancy sensor and a connected lighting load of 0.4 kW, and uses his/her office according to the following schedules, what is the combined lighting load shape for the offices?

Person	Schedule	Percentage of Hour
Joe	8:00 A.M. – 4:00 P.M.	50
Bob	9:00 A.M. – 3:00 P.M.	80
Frank	7:00 A.M. – 11:00 A.M.	100
	2:00 P.M. – 7:00 P.M.	60
Cindy	8:00 A.M. – 7:00 P.M.	100
Oliver	9:00 A.M. – 6:00 P.M.	20

I prefer to calculate the load by taking the percent of each office that is occupied during each hour times the connected load for each office, and summing these (e.g., at 7:00 A.M. the total load would be 0.4 kW since only one person at 100 percent with a connected load of 0.4 kW is in the office). Alternatively, you could calculate each office separately and sum the load shapes from each office to get overall lighting load shape. Figure 7-3 presents the lighting load shape based on the occupancy sensor model.

Daylight Sensors

Daylight sensors work by dimming or turning off lights altogether if there are acceptable levels of ambient lighting (e.g., sunshine) in the space. Since daylight sensors require daylight, their use is limited to exterior spaces which have sufficient window area. Although I will not go into detail on how to calculate the daylight entering a space and the reflected daylight within the space, the primary factor which affects whether or not the daylight sensor goes off is whether or not sufficient light (and reflected light) reach the daylight sensor as shown in Figure 7-4. How much solar light

136 Chapter 7

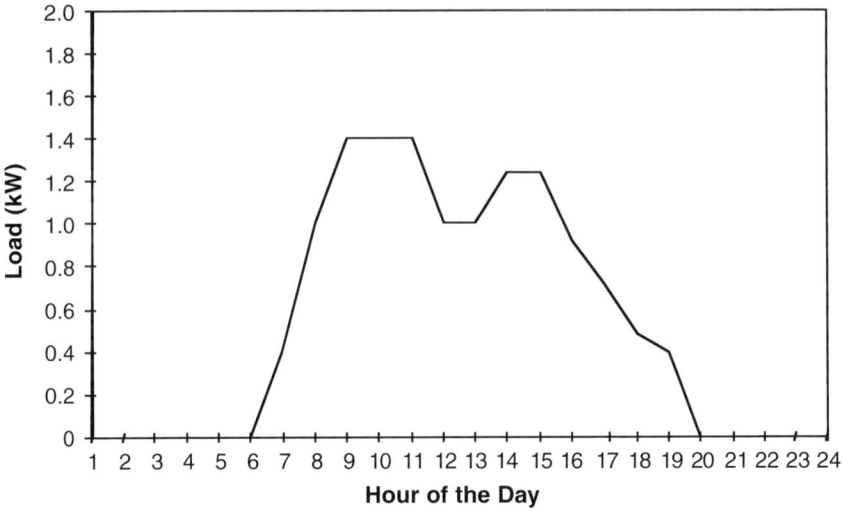

Figure 7-3. Occupancy sensor model.

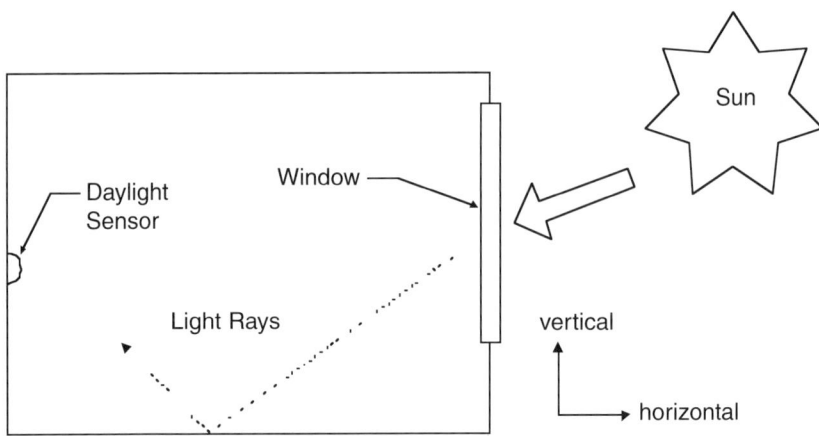

Figure 7-4. Daylight sensors.

enters a space is a function of the time of year, hour of day, orientation, and angle of the sun. Daylight sensors are also commonly used in conjunction with occupancy sensors in exterior offices to provide an even greater level of energy and demand savings. While this book does not provide the details for calculating when occupancy sensors are on, a good reference which provides an introduction to daylight sensor control is the IESNA Lighting Handbook (Illuminating Engineering Society of North America, 1993).

Exterior Lighting

When discussing exterior lights, we are referring to lights attached to the exterior of the buildings (such as security lights), lighted business signs, and parking lot lights. For exterior lights, there are three methods of controlling them: (1) manual switches, (2) timers, and (3) daylight sensors. Whichever control method is utilized, you still model the exterior lighting as a simple box model given the start and stop times. The difference between exterior lighting and indoor lighting is that most exterior lights (such as parking lot lights) go on in the evening and shut off in the early morning, or put another way they have a wrap-around schedule.

If the exterior lights are controlled by a daylight sensor, they turn on at dusk and off at dawn. You can determine when dawn and dusk occur by calculating the sunrise hour and sunset hour for every day of the year given latitudinal and longitudinal coordinates. The hour angle of sunrise and sunset can be calculated using Equation 7-2. The solar equations and tables presented in this section are from data published by Lawrence Berkeley Laboratory, 1983.

$Angle_{srss} = \cos^{-1}[-\tan(Lattitude) \cdot \tan(Declination)]$ **(Equation 7-2)**

where:

$Angle_{srss}$ = hour angle of sunrise and sunset in minutes

Latitude = latitude of geographical location in degrees

Declination = declination angle of sun from Table 7-1 in degrees.

Table 7-1. Solar data for the 21st of each month.

Month	Equation of Time (minimum)	Declination (degree)
January	-11.2	-20.0
February	-13.9	-10.8
March	-7.5	0.0
April	1.1	11.6
May	3.3	20.0
June	-1.4	23.45
July	-6.2	20.6
August	-2.4	12.3
September	7.5	0.0
October	15.4	-10.5
November	13.8	-19.8
December	1.6	-23.45

Next, you can calculate the Apparent Solar Time (AST) for the sunrise and sunset conditions using Equations 7-3 and 7-4, respectively. The AST gives the solar time (in hours) that sunrise and sunset occur.

$$AST_{sunrise} = 12 - \frac{4 \cdot Angle_{srss}}{60} \quad \textbf{(Equation 7-3)}$$

$$AST_{sunset} = 12 + \frac{4 \cdot Angle_{srss}}{60} \quad \textbf{(Equation 7-4)}$$

Finally, you need to translate the apparent solar time into a local standard time (LST) by using Equation 7-5.

$$LST = AST - \frac{ET - 4 \cdot (LSM - Longitude)}{60} + Daylight \quad \textbf{(Equation 7-5)}$$

where:

LST = local standard time

AST = apparent solar time

ET = equation of time, from Table 7-1

Table 7-2. Local standard time meridian (degrees).

Time Zone	LSTM
Atlantic	60
Eastern	75
Central	90
Mountain	105
Pacific	120
Yukon	135
Alaska-Hawaii	150

LSM = local standard time meridian, from Table 7-2

longitude = longitude of geographical location

daylight = daylight savings flag (0 if not daylight savings, 1 otherwise).

If you want to be able to calculate the sunrise and sunset hours on a daily basis, instead of using values for the 21st of each month as shown in Table 7-1, you may use Equation 7-6 and Table 7-3 shown below:

$$\left.\begin{matrix} \tan(Declination) \\ ET \end{matrix}\right\} = \begin{bmatrix} A_0 + A_1 \cdot \cos(W) + A_2 \cdot \cos(2W) + A_3 \cdot \cos(3W) + \\ B_1 \cdot \sin(W) + B_2 \cdot \sin(2W) + B_3 \cdot \sin(3W) \end{bmatrix}$$

(Equation 7-6)

where:

W = the day of the year (e.g., 1 to 365) in radians

A_0 through B_3 = coefficients as shown in Table 7-3.

Exterior Lighting Example

Calculate the times that the exterior lights go on and off on July 21 if they are controlled by a photosensor and located in Boston, Massachusetts. First you need to obtain the latitude and longitude for Boston—it is approximately 42 degrees latitude and 71 degrees longitude.

Table 7-3. Coefficients for the solar variables.

Coefficients	tan(*Declination*)	EQ
A_0	-0.00527	$0.696E^{-04}$
A_1	-0.4001	0.00706
A_2	-0.003996	-0.0533
A_3	-0.00424	-0.00157
B_1	0.0672	-0.122
B_2	0.0	-0.156
B_3	0.0	-0.00556

Use Equation 7-2 to find the angle hour of sunrise and sunset:

$$Angle_{srss} = \cos^{-1}[-\tan(42) \bullet \tan(20.6)]$$
$$= -109.8 \text{ minutes}$$

Use Equation 7-3 to calculate the actual solar time of sunrise:

$$AST_{sunrise} = 12 - \frac{4 \bullet 109.8}{60}$$
$$= 4.68 \text{ hours}$$
$$= 4:48 \text{ A.M.}$$

Use Equation 7-4 to calculate the actual solar time of sunset:

$$AST_{sunset} = 12 + \frac{4 \bullet 109.8}{60}$$
$$= 19.32 \text{ hours}$$
$$= 7:19 \text{ P.M.}$$

Next, use Equation 7-5 to convert the sunrise hour into local time:

$$LST = 4.68 - \frac{6.2 - 4 \bullet (75 - 71)}{60} + 1$$
$$= 5.31 \text{ hours}$$
$$= 5:19 \text{ A.M.}$$

Finally, use Equation 7-5 to convert the sunset hour into local time:

$$LST = 19.32 - 6.2 - \frac{4 \cdot (75 - 71)}{60} + 1$$
$$= 19.95 \text{ hours}$$
$$= 7:57 \text{ P.M.}$$

Miscellaneous Equipment

Miscellaneous equipment is really a catch-all category which can contain a mix of things ranging from motors and compressors to soda machines. Generally things in this category are small pieces of equipment that are difficult to categorize into another end use definition, with the possible exception of computer equipment which is commonly placed in this category. It is also important to realize that the equipment which makes up the end-use categories can change from one sector to another. For example, motors might be ignored when inventorying residential end uses, grouped into the miscellaneous end use during the survey of a commercial building, and placed in the motors end use for an industrial study due to the prevalence of motors in industrial facilities.

Since miscellaneous equipment is such a hodgepodge of equipment types, it can best be estimated using simpler models, such as the box model. If you are modeling equipment which cycles on and off with regularity then you could use an on-off, wave, or sinusoidal model.

Box Model

The box model for miscellaneous equipment is similar to the box model for indoor lighting except that it has an added parameter—a diversity factor. Whereas with indoor lighting we were concerned with the connected load, most miscellaneous equipment operates at part-load conditions which are a fraction of its connected load. The diversity factor is the ratio of actual load over connected load over the time period that the equipment is turned on. The load at any of the on hours for the equipment is equal to the connected load times the diversity factor as presented in Equation 7-7.

$$Load_i = Load_{connected} \cdot Diversity\ Factor \quad \textbf{(Equation 7-7)}$$

where:

$Load_i$ = actual load when the equipment is on

$Load_{connected}$ = nameplate rating of the equipment

$Diversity\ Factor$ = percent of rated load which is drawn when the equipment is on.

Miscellaneous Equipment Example 7-3: Box Model

Given an industrial process which operates from 6:00 A.M. to 10:00 P.M., has a connected load of 30 kW, and a diversity factor of 80 percent, what is the typical load shape using a box model? First, calculate the actual load using Equation 7-7 as shown below:

$$Load_i = 30\ kW \cdot 0.80$$
$$= 24\ kW$$

Now you can generate the load shape given the start and stop times and actual load as shown in Figure 7-5.

On-Off Model

The on-off model for miscellaneous equipment is the first of three predictive models that can be used to estimate process load shapes

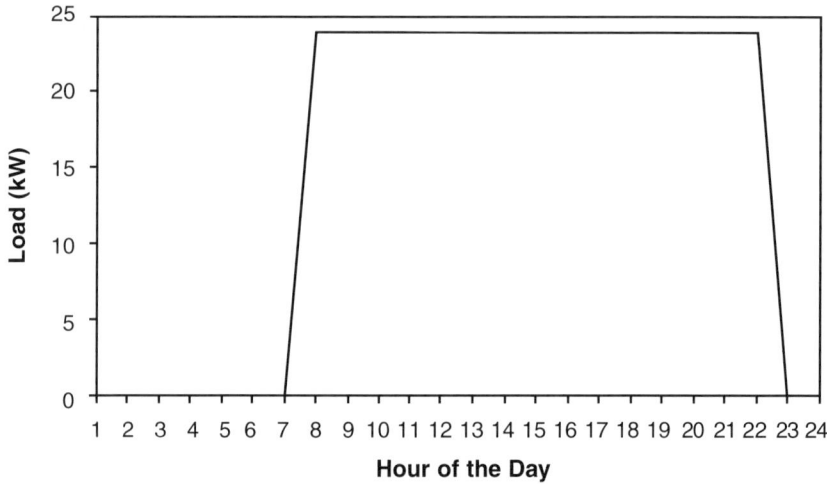

Figure 7-5. Industrial process box model.

Engineering Analysis

(or other cyclic loads). Whereas the box model ignores cycling, on-off models are better at predicting the behavior of cyclic equipment. The on-off model requires five parameters:

1. actual peak load of equipment
2. start time of equipment during the day
3. stop time of equipment during the day
4. time equipment is on
5. time period when equipment is off

The actual peak load of the equipment is calculated by multiplying the connected load of the equipment by the peak load diversity factor as shown in Equation 7-8.

$$Load_{actual} = Load_{connected} \cdot Diversity\ Factor_{peak\ load} \quad \textbf{(Equation 7-8)}$$

where:

$Load_{actual}$ = actual load when the equipment is on

$Load_{connected}$ = nameplate rating of the equipment

Diversity Factor = percent of rated load which is drawn when the equipment is on.

Miscellaneous Equipment Example 7-4: On-Off Model

Given an industrial process which operates from 8:00 A.M. to 9:00 P.M., has a connected load of 30 kW, and a diversity factor of 95 percent, a cycling on time of three hours and a cycling off time of two hours, what is the typical load shape using the on-off model? First, calculate the actual load using Equation 7-8 as shown below:

$$Load_i = 30\ kW \cdot 0.95$$
$$= 28.5\ kW$$

Now you can generate the load shape given the start and stop times and actual load as shown in Figure 7-6.

Sawtooth Model

The sawtooth equipment model uses the same parameters as the on-off model but instead of staying on during the entire on cycle,

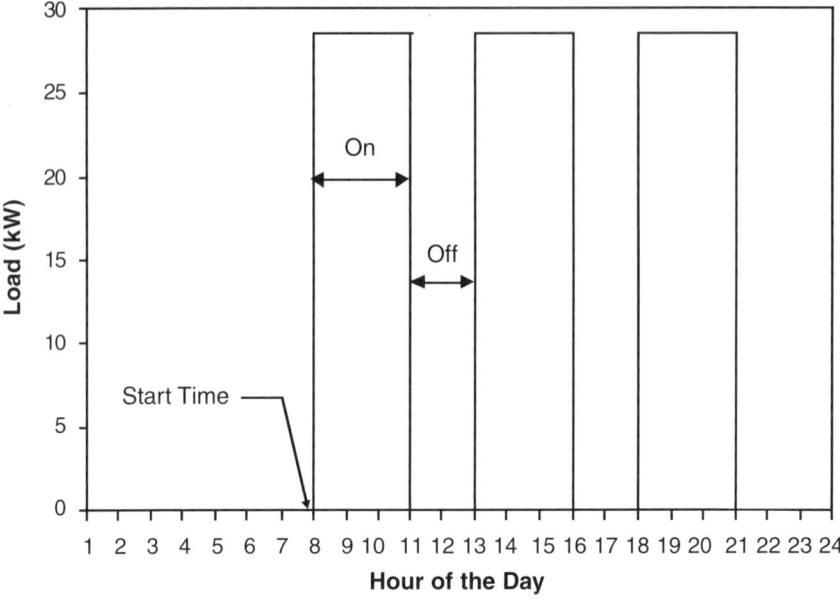

Figure 7-6. Industrial process on-off model.

the sawtooth model starts at the actual load and ramps down to zero load at the end of the on cycle.

Miscellaneous Equipment Example 7-5: Sawtooth Model

Given an industrial process which operates from 6:00 A.M. to 9:00 P.M., has a connected load of 30 kW, and a diversity factor of 95 percent, a cycling on time of four hours and a cycling off time of one hour, what is the typical load shape using the on-off model? First, calculate the actual load using Equation 7-8 as shown below:

$$Load_i = 30 \text{ kW} \cdot 0.95$$
$$= 28.5 \text{ kW}$$

Now you can generate the load shape given the start and stop times and actual load as shown in Figure 7-7.

Sinusoidal Model

The sinusoidal equipment model uses a sine curve to model the equipment load during the day. To use the sinusoidal you need the following parameters:

Engineering Analysis

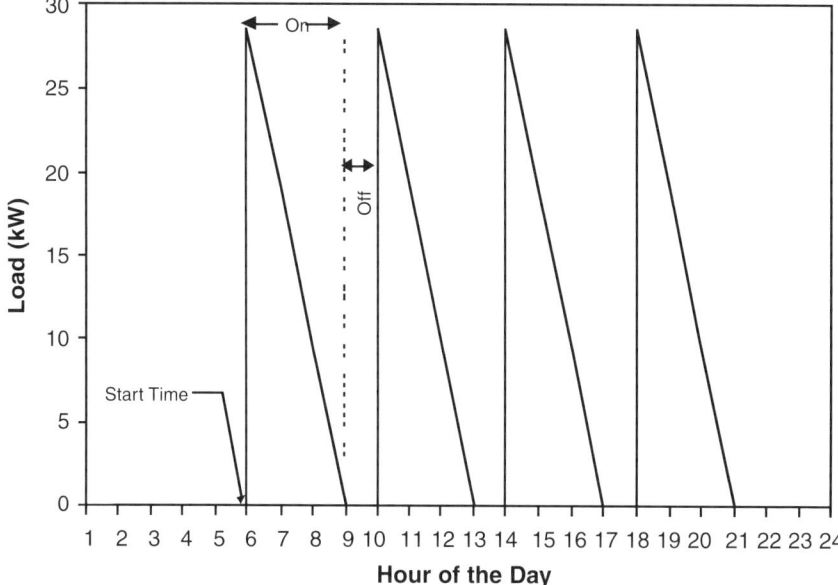

Figure 7-7. Industrial process sawtooth model.

- peak load of equipment
- peak-to-peak cycling time

Miscellaneous Equipment Example 7-6: Sinusoidal Model

Given an industrial process which operates from 7:00 A.M. to 7:00 P.M., has a connected load of 30 kW, and a diversity factor of 95 percent, and a peak-to-peak cycling time of four hours, what is the typical load shape using the sinusoidal model? As with Examples 7-4 and 7-5, the actual peak load is 28.5 kW.

Now you can generate the load shape given the start and stop times and actual load as shown in Figure 7-8.

Domestic Hot Water

In residences, domestic hot water provides for most or all of the hot water needs in the home. In commercial and industrial facilities hot water is also used in different proportions and for different applications. It is important to realize that in addition to heating hot water

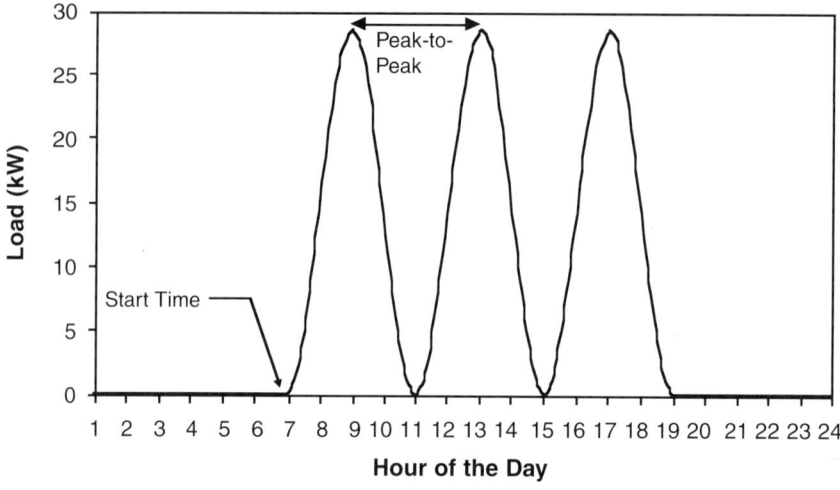

Figure 7-8. Industrial process sinusoidal model.

when it is actively required, the hot water is still being heated during off-peak times to maintain a steady temperature—this is known as standby losses. Three simple models will be presented for modeling annual residential hot water energy use along with statistics on residential standby losses (Pratt and Ross, 1991). Additionally, one model will be presented for commercial buildings.

Linear Model

The first hot water model is the linear model, which predicts the annual hot water energy as a function of the number of occupants in a residence as shown in Equation (7-9).

$$Energy = a + b \cdot No.\ Occupants \qquad \textbf{(Equation 7-9)}$$

where

Energy = annual hot water energy use

a = intercept (in annual kWh)

b = coefficient (in annual kWh per occupant)

No. Occupants = number of occupants in the residence.

Based upon a study in the Pacific Northwest the coefficients for a and b are presented in Table 7-4.

Table 7-4. Linear hot water model results.

Variable	Coefficient	T Value
a	1,917	8.1
b	864	12.3

Square Root Model

The second hot water model is the square root model, which predicts the annual hot water energy as a function of the square root of the number of occupants in a residence as shown in Equation 7-10.

$$Energy = a + b \cdot \sqrt{No.\ Occupants}$$ **(Equation 7-10)**

where:
 $Energy$ = annual hot water energy use
 a = intercept (in annual kWh)
 b = coefficient (in annual kWh per occupant)
 $No.\ Occupants$ = number of occupants in the residence.

Based upon a study in the Pacific Northwest the coefficients for a and b are presented in Table 7-5.

Note that results indicate that if you have no people in the home, the annual hot water energy use would be negative, which is not possible. Therefore, this models requires at least one person in the home.

Table 7-5. Square root hot water model results.

Variable	Coefficient	T Value
a	-967	-2.1
b	3,242	12.5

Age Group Model

The third hot water model is the age group root model, which predicts the annual hot water energy as a function of the square root of the ages of the occupants in a residence as shown in Equation 7-11.

$$\begin{aligned}Energy = {} & a + b \text{ (head of household} > 65) + \\ & c \text{ (No. Occupants} < 6) + \\ & d \text{ (No. Occupants 6 to 17)} + \\ & e \text{ (No. Occupants 18 to 65)} + \\ & f \text{ (No. Occupants} > 65)\end{aligned}$$

(Equation 7-11)

where:

$Energy$ = annual hot water energy use

a = intercept (in annual kWh)

b through f = coefficients (in annual kWh per occupant).

Based upon a study in the Pacific Northwest the coefficients for a through f are presented in Table 7-6.

Standby Losses

While the hot water load models presented above already account for standby losses in residences, the standby losses themselves appear to be different based upon the location of the hot water heat in the home. The average standby losses for water heaters in a study done in the Pacific Northwest were about 1,200 kWh per year.

Table 7-6. Age group hot water model results.

Variable	Coefficient	T Value
a	2,968	11.5
b	-535	-1.4
c	361	2.1
d	901	8.5
e	951	4.9
f	489	1.3

Water Consumption Model

While the previous three models give an indication of the hot water use in residential buildings, for commercial buildings we will examine an engineering model. The most basic hot water model takes into account the quantity of water heated, the inlet temperature of the water to the heater, and the temperature to which the water is heated as presented in Equation 7-12.

$$BTU = 8.34 \cdot Gallons \cdot [T_i - T_o]$$ **(Equation 7-12)**

where:

8.34 = energy to heat 1 gallon of water in BTU

BTU = energy to heat water in BTU

$gallons$ = quantity of water heated in gallons

T_i = hot water inlet temperature in F

T_o = hot water outlet temperature in F.

Water Consumption Model Example

Given an office building which is located in Texas has a water inlet temperature of 55 degrees F throughout the entire year and uses hot water when occupied according to the following schedule:

Hour	Water (Gallons)	Hour	Water (Gallons)	Hour	Water (Gallons)
1	0	9	24	17	14
2	0	10	22	18	11
3	0	11	23	19	5
4	0	12	32	20	2
5	0	13	28	21	0
6	0	14	25	22	0
7	12	15	22	23	0
8	18	16	19	24	0

Additionally, the gas hot water tank is 85 percent efficient, is rated for producing 25,000 BTU/hr of hot water, has standby losses of 1,500 BTU/hr when not utilized and no standby losses when running at the rated conditions, and has a tank temperature setpoint of 120 degrees F. What is the typical hot water load shape (in BTUs) for this facility?

The first step is to calculate the BTUs required to heat the water from 55 degrees F to 120 degrees F at every hour. Second, we need to calculate the standby losses for every hour of the day using an approximation of standby losses during any hour, such as Equation 7-13.

$$\text{Standby Loss} = \text{Loss}_{no\ load} \cdot \left[1 - \frac{\text{Usage}_h}{\text{Usage}_{rated}}\right] \quad \textbf{(Equation 7-13)}$$

where:

Standby Loss = standby loss during hour h

$\text{Loss}_{no\ load}$ = losses when the tank isn't being used

Usage_h = BTU of water heated during hour h

Usage_{rated} = rated BTU of water that can be heated during hour h.

The final step is to convert the BTUs to heat the water into the amount of fuel required to heat the water by dividing them by the efficiency (0.85). The results are summarized in Table 7-7.

Cooking

The cooking end use encompasses all of the appliances which are used to cook meals in residential and commercial applications. Three methods can be used to account for the energy required to cook food are the four-mode, two-mode, and meal-intensity cooking models.

Four-Mode Cooking Model

The four-mode cooking model uses the number of hours that an appliance operates at different loading conditions to determine the overall energy use of the appliance during a typical day. The four different loading conditions are idle, light, medium, and heavy as defined by ASTM Cooking Performance standards for different types of appliances. Equation 7-14 can be used to calculate the daily energy use of an appliance.

Table 7-7. Hot water consumption model results.

Hour	Hot Water (BTU)	Standby Losses (BTU)	Total Btu Required	Fuel Input (MBtu)
1	0	1,500	1,500	1.8
2	0	1,500	1,500	1.8
3	0	1,500	1,500	1.8
4	0	1,500	1,500	1.8
5	0	1,500	1,500	1.8
6	0	1,500	1,500	1.8
7	6,497	1,110	7,608	9.0
8	9,746	915	10,661	12.5
9	12,995	720	13,715	16.1
10	11,912	785	12,697	14.9
11	12,453	753	13,206	15.5
12	17,326	460	17,787	20.9
13	15,161	590	15,751	18.5
14	13,536	688	14,224	16.7
15	11,912	785	12,697	14.9
16	10,288	883	11,170	13.1
17	7,580	1,045	8,625	10.1
18	5,956	1,143	7,099	8.4
19	2,707	1,338	4,045	4.8
20	1,083	1,435	2,518	3.0
21	0	1,500	1,500	1.8
22	0	1,500	1,500	1.8
23	0	1,500	1,500	1.8
24	0	1,500	1,500	1.8

$$Use_{daily} = \sum_{m} Use_m \cdot Hours_m \quad \text{(Equation 7-14)}$$

where:

Use_{daily} = energy use during the day (in either MBtu or kWh)

Use_m = energy use of appliance at load condition m (e.g., light, heavy, etc.)

$Hours_m$ = hours that the appliance operations at condition m.

Alternatively, the appliance load at any hour can be estimated by assuming the loading condition at the specified hour and using the energy use at that condition as a surrogate for the energy use of the appliance.

Four-Mode Cooking Example

Given a convection oven which operates four hours at idle, two hours at light load, two hours at medium load, and one hour at heavy load, what is the daily energy use of the oven? Assume that at the different loading conditions, the oven requires the following fuel input:

Load	Fuel Input (MBtu/hr)
Idle	20
Light	40
Medium	60
Heavy	80

The daily energy use is calculated using Equation 7-14 as shown below:

$$Use_{daily} = 20 \cdot 4 + 40 \cdot 2 + 60 \cdot 2 + 80 \cdot 1$$
$$= 360 \text{ MBtu}$$

Two-Mode Cooking Model

The two-mode cooking model uses the quantity of food cooked and the cooking efficiency of the appliance to estimate the daily energy use of the appliance. Equations 7-15a through 7-15c can be used to calculate the daily energy use of an appliance.

$$Use_{daily} = Use_{Full} \cdot Hours_{Full} + Use_{Idle} \cdot Hours_{Idle} \quad \textbf{(Equation 7-15a)}$$

where:

Use_{daily} = daily energy use of appliance

Use_{full} = fuel input when appliance operates at full load

$Hours_{full}$ = hours appliance operates at full load (see Equation 7-15b)

Use_{idle} = fuel input when appliance operates at idle mode

$hours_{idle}$ = hours appliance operates at idle load, (see Equation 7-15c).

$$Hours_{Full} = \frac{Food\ Cooked}{Cooking\ Efficiency}$$ **(Equation 7-15b)**

where:

Food Cooked = pounds of food cooked during the day

Cooking Efficiency = pounds of food per hour cooked when appliance is at full load.

$$Hours_{Idle} = Hours_{Daily} - Hours_{Full}$$ **(Equation 7-15c)**

where:

$Hours_{Daily}$ = total hours per day that appliance is on

$Hours_{Full}$ = hours that appliance operates at full load.

Two-Mode Cooking Example

Given a charbroiler that operates 12 hours per day, cooks 120 pound of burgers per day, and has a cooking efficiency of 45 pounds per hour, what is the daily energy use of the appliance? The fuel input of the appliance is 60 MBtu/hr at idle load and 80 MBTU/hr at full load.

Step 1: Using Equation 7-15b then equivalent full load hours for the charbroiler are:

$$Hours_{Full} = \frac{120\ lb}{45\ lb/hr}$$
$$= 2.7\ hr$$

Step 2: Calculate the idle hours using Equation 7-15c:

$$Hours_{Idle} = 12 - 2.7$$
$$= 9.3\ hr$$

Step 3. Calculate the average daily energy use of the appliance:

$$Use_{daily} = 80\ \frac{MBtu}{hr} \cdot 2.7\ hr + 60\ \frac{MBtu}{hr} \cdot 9.3\ hr$$
$$= 774\ MBtu$$

Meal Intensity Model

The meal intensity model bases the energy use on the number of meals served rather than examining the appliances in the kitchen. The additional information required is how heavily the appliances in the kitchen are used for each of the primary meals—breakfast, lunch, and dinner—as well as the hours that those meals occur.

One estimate of the amount of energy required to cook the meals for different types are restaurants is listed below:

Restaurant Type	BTU's per Meal
Full service restaurant	4,000
Cafeteria	4,000
Fast food	3,000

Given the hours that each meal (breakfast, lunch, and dinner) occurs, you can use a normal distribution to distribute the cooking energy use over a time period for each meal. First, you need to determine the total energy use for cooking the meal as shown in Equation 7-15a.

$$Energy_{meal} = \frac{Btu's}{Meal} \cdot No.\ Meals \quad \textbf{(Equation 7-16a)}$$

where:

$Energy_{meal}$ = energy required to cook all of the meals during a given period (breakfast, lunch, or dinner)

$BTU's/meal$ = energy to cook a single meal, based upon the restaurant type

$No.\ Meals$ = number of meals served during a given period.

Next, you distribute that total energy, centered around the mid-meal hour, using a distribution scheme to determine the cooking energy required during each hour of cooking. You could either use a normal distribution to distribute the total cooking energy over each hour of the meal or you could use a functions such as those presented in Equations 7-16b and 7-16c.

$$Energy_h = \frac{Fraction_h}{\sum Fraction_h} \cdot Energy_{meal} \quad \textbf{(Equation 7-16b)}$$

where:

$Energy_h$ = energy required for cooking at hour h

$Fraction_h$ = fraction of energy used to cook during hour h of the meal period (see Equation 7-16c).

$$Fraction_h = \frac{1}{|h - h_{mid}| + Duration} \quad \textbf{(Equation 7-16c)}$$

where:

h = specific hour in the meal period

h_{mid} = fractional midpoint hour of the meal time (e.g., if the meal was served from 6:00 A.M. to 9:00 A.M., the midpoint hour would be 7.5)

Duration is the length of the meal in hours.

Meal Intensity Model Example 1

Given that 2,000 meals are served for breakfast (6:00 A.M. to 9:00 A.M.) in a fast food restaurant, what is the cooking load shape for those hours?

Step 1: Calculate the total cooking energy required to cook the meals, using Equation 7-16a:

$$Energy_{meal} = 3,000 \frac{Btu's}{Meal} \cdot 2,000 \; Meals$$
$$= 6,000 \; MBtu$$

Step 2: Calculate the fraction of food that is cooked during each breakfast hour using Equation 7-16c. The midpoint hour is 7.5 and the duration of the breakfast period is three hours.

Step 3: Calculate the Btu's required during each hour of the cooking period by using Equation 7-16b. The results are summarized on the following page:

Hour	Fraction$_h$	MBTU
6	0.22	1,313
7	0.29	1,688
8	0.29	1,688
9	0.22	1,313

Meal Intensity Model Example 2

Given the following meal schedule in a full service restaurant, what is the cooking load shape?

Meal	Meals Served	Time Served
Breakfast	1,500	6:00 A.M. to 10:00 A.M.
Lunch	2,500	11:00 A.M. to 1:00 P.M.
Dinner	3,000	4:00 P.M. to 8:00 P.M.

Step 1: Calculate the total MBTU for each of the meals.

Meal	Total MBTU
Breakfast	6,000
Lunch	10,000
Dinner	12,000

Step 2: Calculate the fraction of meals served each hour.

Step 3: Calculate the MBTU used each hour for cooking. The results are presented as Figure 7-9.

Refrigeration/Freezers

Refrigerators and freezers are one of the major energy consuming end uses in grocery stores, convenience stores, and similar retail establishments. The energy use of refrigeration and freezer cases depend upon several factors including: type of case, how many people are in the store, and whether or not the cases have automatic defrost.

Since refrigeration and freezer cases have a relatively constant load throughout the year, one method of determining the load at

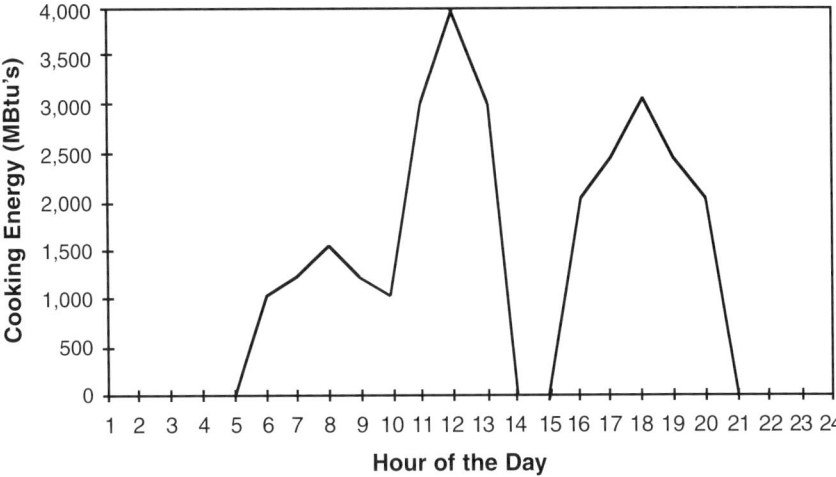

Figure 7-9. Cooking example 2 load shape.

any hour is to simply divide the annual energy use by 8,760 (the number of hours in a year). The primary types of refrigeration/freezer cases are:

- open coffin
- covered coffin
- open multi-deck
- closed multi-deck
- self-contained
- walk-in

Refrigeration Energy Use Model

The annual energy use for different case types can be calculated from the equations contained in Table 7-8. Whether the freezer is operating in refrigeration or freezer mode is dependent solely on the temperature of the case. In general, if the food is being stored at 32 degrees F or below, the case is in freezer mode.

Table 7-8. Refrigeration and freezer annual energy use equations.

Case Type	Refrigeration Mode (kWh)	Freezer Mode (kWh)
Open coffin	650 • Length	800 • Length
Covered coffin	600 • Length	800 • Length
Open multi-deck	2,200 • Length	3,400 • Length
Closed multi-deck	1,300 • Length	2,600 • Length
Self-contained	1,200 • Length	1,650 • Length
Walk-in	2,700 + 100 • Area	3,200 + 200 • Area

Once you know the annual energy use of the refrigeration/freezer equipment, you can calculate the end-use demand at any hour using Equation 7-17.

$$Demand = \frac{Annual\ kWh}{8,760}$$ **(Equation 7-17)**

Additional Refrigeration Factors

Two additional factors that you may wish to account for when modeling refrigeration systems are the effects of occupancy and the presence of electric defrost for the cases. During occupied periods in a store, the refrigeration and freezer cases generally consume about 10 percent more energy than during unoccupied periods. This is explained by the fact that as people remove produce, dairy products, or other items from the cases throughout the day, the cases are restocked with additional items that need to be cooled to the appropriate temperature. Electric defrost on refrigeration and freezer cases adds about 5 percent additional energy use on average to the energy use of the case.

Refrigeration Effect

Refrigerators and freezers have another unique effect on the store; they actually help to cool and dehumidify the air in the store. Table 7-9 presents the amount of cooling that the cases provide by case type. In effect, the cases provide a negative heat gain to the space.

Table 7-9. Refrigeration effect from cases.

Case Type	Refrigeration (BTU/hr-ft)	Freezers (BTU/hr-ft)
Open coffin	-150	-250
Covered coffin	0	0
Open multi-deck	-620	-980
Closed multi-deck	0	0
Self-contained	-150	-250
Walk-in	0[a]	0[a]

[a] The units for walk-in case are BTU per hr-ft^2

Ventilation

The primary ventilation system is used in conjunction with the heating and cooling systems to deliver the conditioned air to the appropriate space. In order to properly model the ventilation load you need to know the size of motors on the air handler fans in the systems. A rough approximation of the ratio of air handler CFM to motor is given in Table 7-10.

Heating and Air Conditioning

Due to their complexity and excellent coverage in other publications, I won't provide detailed information on how to calculate heat-

Table 7-10. Ratio of air handler CFM to motor hp.

Ventilation System Type	$\dfrac{CFM}{hp}$
Package system Split system Air handler	2,300
Furnace Unit ventilator Window unit	2,700
Variable air volume	800

ing and air-conditioning loads. Instead, please refer to calculation methods given in the *ASHRAE Fundamentals* volume or another reference book. Alternatively, you could use the load shapes you have developed for the other end uses and use them as input into a more complicated building simulation model, such as DOE-2.

The design heating load in a building is governed by several factors, including:

- heat loss through roofs, walls, and windows
- heat loss through basements walls
- heat loss through floors and slabs
- infiltration and ventilation air

The factors are shown schematically in Figure 7-10.

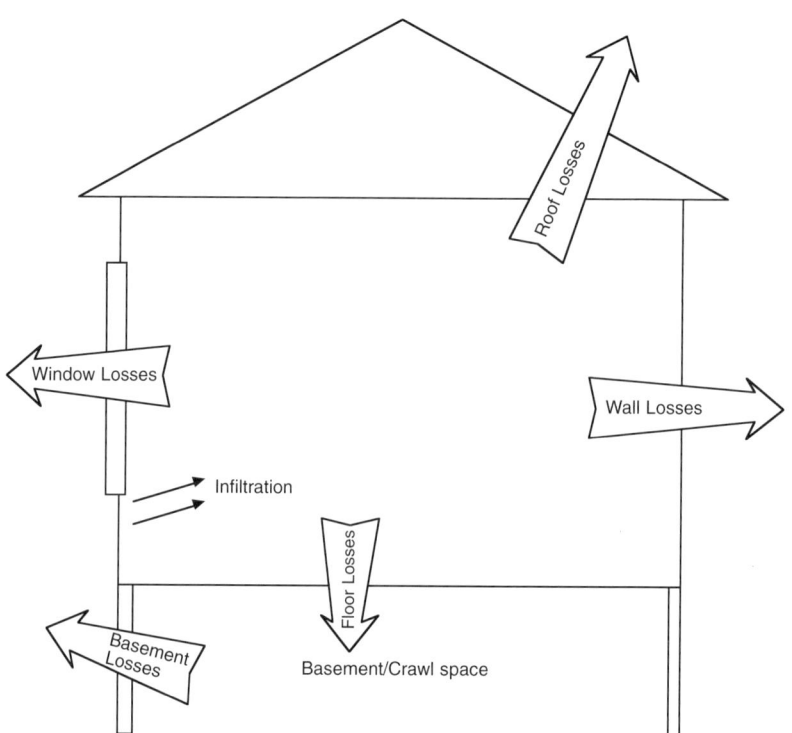

Figure 7-10. Design heating load schematic.

Air Conditioning

The design air-conditioning load in a building also has several contributing factors, including:

- heat gain through roofs and walls
- conduction through windows
- solar gain through windows
- gain through partitions and floors
- internal gains from lights
- internal gains from occupants
- internal gains from appliances
- ventilation and infiltration gains

These factors are shown schematically as Figure 7-11.

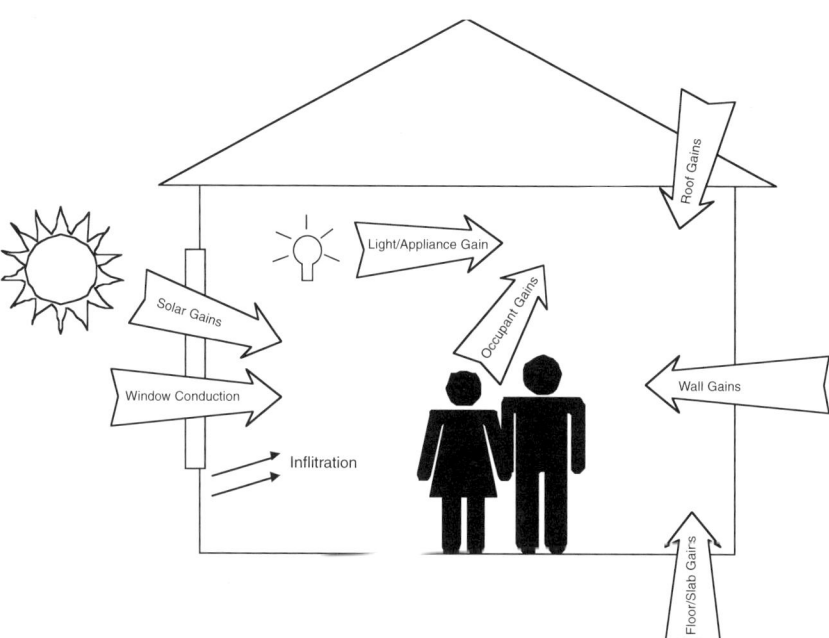

Figure 7-11. Design cooling load schematic.

Load Shape Aggregation

Once you have developed engineering estimates of load shapes, the aggregation techniques are the same as those presented in Chapter 6. However, if you have either the annual energy use of the facility or whole-building metered data you may wish to do a simple calibration of the engineering estimates to the energy use or hourly demand loads.

Calibrating Load Shapes

Calibrating your load shape estimates serves a couple of purposes. This first purpose is that it provides a level of quality control since you are comparing an estimated value to the actual value. The second purpose is that if you are using the load shapes for planning or forecasting, you ensure that the estimated population energy or hourly demand match the true values. Two straightforward methods of calibrating estimated loads to energy use are the annual energy calibration and monthly energy calibration methods. I will also present a simple method for calibrating engineering estimates to hourly load data.

Annual Energy Calibration

In the annual energy calibration method, you need to know the annual energy use of the customer, or population, that you are estimating. To use the annual method you have two steps: (1) calculate the calibration factor based on the annual data, and (2) apply that calibration factor to all of your end-use load shape estimates.

The annual calibration factor can be calculated using Equation 7-18 shown below:

$$R = \frac{\text{Actual Annual Energy Use}}{\text{Estimated Annual Energy Use}} \quad \textbf{(Equation 7-18)}$$

where:

R = calibration factor.

Next, you apply the calibration factor to every daytype and hour of each end-use load shape estimate using Equation 7-19.

$$EU_{Cal,h,d} = EU_{Est,h,d} \cdot R \qquad \textbf{(Equation 7-19)}$$

where:

$EU_{Cal,h,d}$ = calibrated end-use load for hour h and daytype d

$EU_{Est,h,d}$ = estimated end-use load for hour h and daytype d.

If you are using the calibration ratio to perform quality control checks on an individual site, R should be between 0.75 and 1.25. If you are outside of this range, chances are good that either you have the wrong or incomplete billing history data or that your engineering estimates are faulty somewhere.

Annual Energy Use Calibration Example

Assume that you have an office building with an annual billing history of 25,000 kWh and an estimated annual energy use of 30,000 kWh. What is the calibration ratio?

Using Equation 7-18, the calibration ratio is:

$$R = \frac{25,000 \text{ kWh}}{30,000 \text{ kWh}}$$
$$= 0.83$$

Monthly Energy Calibration

If you want a little bit more control over the calibration and you have access to monthly billing histories for your customers or the population, you can use the monthly energy calibration method. The steps are the same as those for the annual energy calibration method except the you analyze each month individually—calculating a calibration ratio and applying it to the appropriate daytypes for the specific month.

Hourly Load Calibration

For the hourly load calibration, you are even going to a higher level of calibration where you calculate a calibration factor for each hour of every daytype and apply the ratio to the end uses for those hours and daytypes. Instead of using energy use, however, you calculate the calibration ratio using demand (kW) for electricity or BTUs for natural gas as shown in Equation 7-20.

$$R = \frac{\text{Actual Hourly Demand}}{\text{Estimated Hourly Demand}}$$ **(Equation 7-20)**

Hourly Load Calibration Example

Given the following hourly values for both the whole-building load research data and the estimated sum of the end uses for a facility, what is the calibration ratio for each hour?

Hour	Estimate (kW)	Load Data (kW)	Calibration Ratio
1	7	7	1.00
2	6	4	0.67
3	5	2	0.40
4	6	6	1.00
5	8	6	0.75
6	10	11	1.10
7	15	16	1.07
8	22	18	0.82
9	25	24	0.96
10	24	27	1.13
11	27	32	1.19
12	25	27	1.08
13	30	32	1.07
14	28	24	0.86
15	29	30	1.03
16	26	15	0.58
17	22	19	0.86
18	17	13	0.76
19	12	9	0.75
20	8	7	0.88
21	4	3	0.75
22	5	5	1.00
23	6	6	1.00
24	7	1	0.14

One of the things that you should get out of this chapter is that self-generated engineering models provide a lot more flexibility and customization than "canned" building simulation software. In Chapter 8, statistical models and methods for developing end-use load shapes will be introduced.

References

"DOE-2 Engineer's Manual," LBL-11353, Lawrence Berkeley Laboratory, 1983.

"IESNA Lighting Handbook-8th Edition," HB-93, Illuminating Engineering Society of North America, 1993.

McQuiston, Faye C., and Spitler, Jeffrey D., "Cooling and Heating Load Calculation Manual 2nd Edition," ASHRAE, 1992.

Pratt, R.G., and Ross, B.A., "Measured Electric Hot Water Standby and Demand Loads from Pacific Northwest Homes," PNL-7889, Pacific Northwest Laboratory, November 1991.

8
Statistical Analysis

Statistical Approaches

There are a wide variety of statistical techniques that can be used to develop end-use load shapes. One of the more common techniques is Conditional Demand Analysis (CDA), used to disaggregate billing history into end-use components. Additionally, CDA can be used to disaggregate whole-building load research data into end-use load shape estimates.

Statistical analysis can also be utilized in conjunction with engineering models to help refine end-use estimates. Two commonly used methods which combine statistical and engineering analysis are the Statistically Adjusted Engineering (SAE) model and the Hybrid Statistical Engineering Method (HSEM). While the SAE model directly applies the results of statistical analysis to refine the end-use load shapes, the HSEM approach uses the results of the statistical analysis to identify areas where the engineering models can be improved.

Annual Conditional Demand Analysis

CDA is probably the backbone of the statistical methods used to develop end-use load shapes. Some of the earliest applications of CDA were to disaggregate annual billing history for a population of residences into average end-use consumption estimates. The basic form of an energy use end-use dissaggregation equation is presented as Equation 8-1.

$$Energy_f = a_1 \cdot End\text{-}Use_{f1} + a_2 \cdot End\text{-}Use_{f2} + \cdots + a_n \cdot End\text{-}Use_{fn}$$

(Equation 8-1)

where:

$Energy_f$ = annual energy usage for fuel f in each facility

$End\text{-}Use_{fi}$ = flag to indicate the presence of end-use i which uses fuel f in each facility (a 1 indicates the end-use fuel combination exists, 0 indicates the end-use fuel combination is not present)

a_i = resulting regression coefficients for end-use i that represent the amount of energy used by each end use for a group of buildings.

Example 8-1: Annual Energy CDA Example

Assume that you have a group of 30 residential buildings which have the following 10 end uses: hot water, HVAC (heating and air conditioning), lighting, refrigerators, freezers, ranges, clothes dryer, clothes washer, dishwasher, and other. It is also known from their annual billing history that the homes have the annual energy use presented in Table 8-1. Additionally, they have the end-use mix indicated in Table 8-2. Using CDA, estimate the annual energy use of the 10 end uses in each of the homes.

Once the total annual energy use and end-use indicators are known, regression analysis can be performed to determine the magnitude of each end-use component. Using Equation 8-1 we come up with the following equations for house numbers 1, 2, and 30. (I leave the exercise of generating the equations for the remaining houses to the reader.):

$18,055 = a_1 \cdot 1 + a_2 \cdot 1 + a_3 \cdot 1 + a_4 \cdot 1 + a_5 \cdot 0 + a_6 \cdot 1 + a_7 \cdot 1 + a_8 \cdot 0 + a_9 \cdot 0 + a_{10}$

$12,232 = a_1 \cdot 0 + a_2 \cdot 1 + a_3 \cdot 1 + a_4 \cdot 1 + a_5 \cdot 0 + a_6 \cdot 0 + a_7 \cdot 0 + a_8 \cdot 0 + a_9 \cdot 1 + a_{10}$

$\vdots$

$16,805 = a_1 \cdot 0 + a_2 \cdot 0 + a_3 \cdot 1 + a_4 \cdot 1 + a_5 \cdot 1 + a_6 \cdot 0 + a_7 \cdot 1 + a_8 \cdot 1 + a_9 \cdot 1 + a_{10}$

Table 8-1. Annual energy use of residential homes.

House	Annual Electricity Use (kWh)
1	18,055
2	12,232
3	18,195
4	12,295
5	11,450
6	23,450
7	20,951
8	16,457
9	17,100
10	23,627
11	16,440
12	23,524
13	18,510
14	10,824
15	17,382
16	21,369
17	11,912
18	28,446
19	23,501
20	13,536
21	31,265
22	10,703
23	14,528
24	12,335
25	20,877
26	6,530
27	24,868
28	13,394
29	18,953
30	16,805

Table 8-2. End-use mix of residential homes.

House	Hot Water	Heat and Air Conditioning	Light	Refrigeration	Freezer	Range	Clothes Dryer	Clothes Washer	Dish Washer	Other
1	X	X	X	X		X	X			X
2		X	X	X					X	
3		X	X	X	X	X	X	X		
4	X		X						X	X
5	X	X	X	X				X	X	
6	X	X	X	X				X		X
7	X	X	X	X	X	X	X	X	X	X
8			X	X			X	X	X	X
9	X	X	X	X	X					
10	X	X	X	X		X	X		X	
11		X	X	X	X		X	X	X	X
12	X	X	X	X	X	X	X	X	X	X
13	X	X	X	X			X	X	X	X
14			X	X				X		X
15	X	X	X			X		X		X

Statistical Analysis

House	Hot Water	Heat and Air Conditioning	Light	Refrigeration	Freezer	Range	Clothes Dryer	Clothes Washer	Dish Washer	Other
16		X	X		X		X		X	X
17	X	X	X			X	X		X	
18	X	X	X	X		X	X	X	X	X
19	X	X	X			X	X	X	X	X
20	X		X	X	X	X		X		
21	X	X	X	X	X	X	X	X	X	X
22	X		X	X				X	X	
23			X	X	X		X		X	X
24		X	X	X		X		X		
25	X		X	X		X	X	X		X
26		X	X	X			X	X	X	X
27	X	X	X	X	X	X	X	X	X	X
28			X			X	X	X	X	X
29	X	X	X	X		X	X	X	X	
30			X	X	X		X	X	X	X

Table 8-3. Annual energy end-use regression coefficients.

Parameter	End-use	Coefficients	Standard Error	t Stat
a_1	Hot water	3684.9	1527.7	2.41
a_2	HVAC	3697.2	1621.8	2.28
a_3	Lighting	3760.1	2729.3	1.38
a_4	Refrigerators	2460.0	1948.7	1.26
a_5	Freezers	2145.4	1561.7	1.37
a_6	Ranges	2887.0	1854.1	1.56
a_7	Clothes dryer	1527.7	2097.8	0.73
a_8	Clothes washer	322.0	1513.1	0.21
a_9	Dish washer	350.0	1700.2	0.21
a_{10}	Other	5118.3	1710.9	2.99

After obtaining a similar equation for every house, regression analysis can be performed to determine the value of the coefficients (a_1, a_2, etc.). The results of the regression analysis are presented in Table 8-3.

Note that there is not an intercept in the regression equation because it is known that the sum of the annual end-use energy must be equal to the annual energy use for the house. The end-use estimates are equal to the regression coefficient for each end use (e.g., hot water is equal to a_1; light is equal to a_3, etc.).

It's also important to understand that we have just estimated the average end-use consumption across all of the homes, sometimes referred to as cross-sectional analysis. This is probably sufficient if you are trying to determine the average end-use breakdown for a population. Additionally, this technique has historically been applied when a utility is trying to determine the difference between end-use energy for two separate samples (e.g., a control sample and a participant sample), as is commonly the case when performing impact analysis for a utility-sponsored DSM program.

While this technique may do a satisfactory job of determining the end-use energy breakdown for the population, how well did we do estimating the total energy use of each building? Table 8-4

Table 8-4. Comparison of actual annual energy to estimated annual energy.

House	Estimated Total	Actual Total	% Difference
1	23,135	18,055	28.1
2	10,267	12,232	16.1
3	16,800	18,195	7.7
4	12,913	12,295	5.0
5	14,274	11,450	24.7
6	18,721	23,450	20.2
7	25,953	20,951	23.9
8	13,538	16,457	17.7
9	16,070	17,100	6.0
10	18,367	23,627	22.3
11	19,031	16,440	15.8
12	25,953	23,524	10.3
13	20,920	18,510	13.0
14	12,010	10,824	11.0
15	14,351	17,382	17.4
16	19,059	21,369	10.8
17	15,907	11,912	33.5
18	25,953	28,446	8.8
19	21,347	23,501	9.2
20	15,260	13,536	12.7
21	25,953	31,265	17.0
22	10,577	10,703	1.2
23	15,362	14,528	5.7
24	13,126	12,335	6.4
25	20,110	20,877	3.7
26	10,267	6,530	57.2
27	25,953	24,868	4.4
28	13,965	13,394	4.3
29	18,689	18,953	1.4
30	15,684	16,805	6.7

presents the comparison of estimated annual energy use to actual annual energy use for each house. The estimated annual energy use for each house is calculated by adding the product of the regression coefficients for each end use by the end-use indicator.

In the best comparison we were only off by about 1 percent but for one house we overestimated the annual energy use by almost 60 percent and on average we were off by 14 percent. It's important to keep in mind though that the purpose of this analysis was to determine the average end-use energy for the population—not the end-use energy for each individual home.

Monthly CDA Analysis

What happens if we use monthly CDA analysis using the monthly billing history instead of the annual bills for a house? Most of the end uses in a home remain relatively constant across the months while air conditioning and heating energy use can vary significantly by month. Additionally, hot water energy use varies to a lesser extent on a month-by-month basis as a function of ground water temperature (e.g., the colder the supply water is going into the hot water tank, the more energy is required to raise the supply water temperature to the desired setpoint).

If we simply performed a monthly CDA analysis using 0 or 1 indicators to indicate the absence or presence of end uses we would probably obtain very poor results. Therefore, we need to introduce the concept of using proxy parameters in the basic CDA equation. A proxy is a parameter which mimics the change in the variable we are trying to estimate.

One of the simplest proxies to implement when estimating monthly air conditioning energy is cooling degree days and likewise for heating using heating degree days. The TMY2 Users Manual (Marrion and Urban, 1996) defines a degree day as:

> " ... the difference between the average temperature for the day and a base temperature. If the average for the day is less than the base value, then the difference is designated as heating degree days. If the average for the day is greater than the base value, the difference is designated as cooling degree days."

One simple formulation that can be used to disaggregate the monthly energy bills into heating and cooling for a population is shown as Equation 8-2.

$$Energy_m = a_1 \cdot CDD_m + a_2 \cdot HDD_m + a_3 \cdot Other \quad \textbf{(Equation 8-2)}$$

where:

$Energy_m$ = monthly energy use for month m

CDD_m = cooling degree days for month m

HDD_m = heating degree days for month m

$Other$ = indicator of the non-HVAC end uses in the house.

Example 8-2: Monthly Energy CDA Example

We will use the same group of 30 residential homes, used in Example 8-1 to show the monthly energy use (Table 8-5). The heating and cooling degree days for each month are shown in Table 8-6. The final piece of information needed is an indicator for the non-HVAC end uses in the house. In this example, we chose to calculate the other indicator as the sum of the indicators for the non-HVAC end uses (e.g., if a house had five non-HVAC end uses, the other indicator would be five). Table 8-7 presents the other indicators for each house.

Table 8-5. Monthly energy use.

House	Jan.	Feb.	Mar.	Apr.	May	June	July	Aug.	Sep.	Oct.	Nov.	Dec.
1	1,571	1,715	1,513	1,545	1,433	1,342	1,340	1,304	1,296	1,465	1,657	1,875
2	1,431	1,346	1,028	950	924	841	874	824	869	896	982	1,269
3	1,939	1,991	1,557	1,666	1,322	1,177	1,215	1,239	1,200	1,433	1,417	2,039
4	1,025	1,025	1,025	1,025	1,025	1,025	1,025	1,025	1,025	1,025	1,025	1,025
5	1,220	974	1,026	986	796	686	692	730	656	1,016	1,085	1,583
6	2,553	2,352	1,977	2,075	1,690	1,548	1,676	1,642	1,625	1,905	2,055	2,354
7	1,891	2,030	1,774	1,750	1,654	1,581	1,558	1,545	1,583	1,701	1,928	1,958
8	1,371	1,371	1,371	1,371	1,371	1,371	1,371	1,371	1,371	1,371	1,371	1,371
9	1,434	1,629	1,659	1,501	1,303	1,202	1,170	1,189	1,202	1,234	1,698	1,878
10	2,053	2,224	2,525	2,228	1,793	1,592	1,637	1,647	1,632	2,024	2,270	2,002
11	1,751	1,516	1,351	1,291	1,294	1,248	1,187	1,215	1,248	1,336	1,348	1,654
12	2,001	2,127	2,025	2,066	1,918	1,861	1,852	1,816	1,827	1,922	2,052	2,058
13	1,725	1,809	1,670	1,691	1,352	1,250	1,280	1,187	1,292	1,413	1,721	2,119
14	902	902	902	902	902	902	902	902	902	902	902	902
15	1,697	2,046	1,736	1,431	1,316	1,131	995	1,045	1,062	1,182	1,996	1,745

House	Jan.	Feb.	Mar.	Apr.	May	June	July	Aug.	Sep.	Oct.	Nov.	Dec.
16	2,048	2,073	2,114	1,688	1,620	1,502	1,469	1,495	1,565	1,750	1,735	2,311
17	1,039	1,254	1,132	922	785	774	768	772	783	993	1,228	1,461
18	3,145	2,890	2,322	2,579	2,235	1,973	2,046	2,005	1,945	2,216	2,240	2,848
19	2,175	2,007	2,121	2,064	1,799	1,805	1,741	1,768	1,748	1,991	2,197	2,084
20	1,128	1,128	1,128	1,128	1,128	1,128	1,128	1,128	1,128	1,128	1,128	1,128
21	3,466	2,819	2,944	2,443	2,494	2,233	2,257	2,248	2,212	2,352	2,716	3,081
22	892	892	892	892	892	892	892	892	892	892	892	892
23	1,211	1,211	1,211	1,211	1,211	1,211	1,211	1,211	1,211	1,211	1,211	1,211
24	1,121	1,401	1,175	1,266	937	679	720	755	702	1,060	1,101	1,418
25	1,740	1,740	1,740	1,740	1,740	1,740	1,740	1,740	1,740	1,740	1,740	1,740
26	819	596	614	560	450	413	395	404	399	463	635	782
27	2,162	2,121	2,071	2,159	2,084	1,901	1,963	1,977	1,950	2,008	2,231	2,240
28	1,116	1,116	1,116	1,116	1,116	1,116	1,116	1,116	1,116	1,116	1,116	1,116
29	1,903	1,583	1,779	1,681	1,547	1,372	1,389	1,428	1,396	1,586	1,692	1,597
30	1,400	1,400	1,400	1,400	1,400	1,400	1,400	1,400	1,400	1,400	1,400	1,400

Table 8-6. Monthly heating and cooling degree days.

Month	Cooling Degree Days	Heating Degree Days
January	0	809
February	0	610
March	0	592
April	0	438
May	6	263
June	43	118
July	119	35
August	122	51
September	42	111
October	0	332
November	0	585
December	0	747
Annual	332	4,691

Table 8-7. "Other" indicators by house.

House	Other Indicators
1	6
2	3
3	6
4	4
5	5
6	4
7	9
8	6
9	5
10	6
11	6
12	9
13	7
14	5
15	4
16	6

17	5
18	9
19	7
20	6
21	9
22	5
23	6
24	4
25	8
26	3
27	9
28	6
29	7
30	7

We now have all of the information necessary to perform a monthly regression analysis on the homes. The regression results are presented in Table 8-8.

Table 8-8. Monthly regression results.

Parameter	End Use	Coefficients	Standard Error	t Stat
a_1	Heating	0.603	0.083	7.30
a_2	Cooling	0.378	0.526	0.72
a_3	Other	201.9	7.2	28.82

Based on the regression results from Table 8-8, the average estimated annual heating energy for the houses is 2,830 kWh and the average estimated annual cooling energy is about 130 kWh for a total HVAC estimate of about 2,960 kWh annually. This is roughly 20 percent lower estimate than the 3,600 kWh estimated in Example 8-1. It should also be noted that the actual average annual HVAC consumption of the homes is 5,000 kWh which implies that the estimated HVAC energy in Examples 8-1 and 8-2 underestimate the HVAC usage by 30 to 40 percent.

Monthly Time-Series CDA Analysis

In the previous two sections we have shown examples of using annual and monthly cross-sectional analysis to disaggregate the billing history into end-use components. In this section, we will see the effect of performing the monthly billing dissaggregation on individual homes. Recall when we performed regression analysis at a single point in time over several buildings where we were implementing cross-sectional analysis. In this section we will utilize the monthly variance in energy use for a single building to predict end-use energy. This approach is known as time-series analysis.

Example 8-3: Monthly CDA Analysis on Houses 1, 3, and 10

Assume that for house 1 we have the monthly billing history from Table 8-5. Additionally, we have the heating and cooling degree days presented in Table 8-6 and we have determined that the value of the "other" indicator is 6. Applying the data for house 1 to Equation 8-2, we generate the results presented in Table 8-9.

Table 8-9. Regression results for house 1

Parameter	End-Use	Coefficients	Standard Error	t Stat
a_1	Heating	0.410	0.114	3.61
a_2	Cooling	0.083	0.682	0.12
a_3	Other	209.2	10.5	19.86

When examining the t values in Table 8-9, it becomes evident that the cooling end use has a t value which is significantly less than 2, which implies that the parameter is not statistically significant based on a 95-percent confidence level. Given the regression results, the monthly energy use estimates for house 1 are shown below:

Month	Heating	Cooling	Other	Total
January	332	—	1,255	1,587
February	250	—	1,255	1,506
March	243	—	1,255	1,498
April	180	—	1,255	1,435
May	108	0	1,255	1,364
June	48	4	1,255	1,307
July	14	10	1,255	1,280
August	21	10	1,255	1,287
September	46	3	1,255	1,304
October	136	—	1,255	1,392
November	240	—	1,255	1,495
December	306	—	1,255	1,562
Total	1,924	28	15,066	17,017

Given that the actual HVAC usage for house 1 is about 3,500 kWh annually, it is clear that the regression underestimated the results for this particular house. Next, let's take a look at the regression results for House 3, as shown in Table 8-10.

Table 8-10. Regression results for house 3.

Parameter	End Use	Coefficients	Standard Error	t Stat
a_1	Heating	1.21	0.26	4.63
a_2	Cooling	1.37	1.57	0.88
a_3	Other	167.5	24.2	6.92

Once again, the t values in Table 8-10 for cooling are still less than 2. Using the regression results, the end-use estimates were calculated and are listed on the following page:

Month	Heating	Cooling	Other	Total
January	980	—	1,005	1,985
February	739	—	1,005	1,744
March	717	—	1,005	1,722
April	531	—	1,005	1,535
May	319	8	1,005	1,332
June	143	59	1,005	1,207
July	42	163	1,005	1,210
August	62	167	1,005	1,234
September	134	58	1,005	1,197
October	402	—	1,005	1,407
November	709	—	1,005	1,714
December	905	—	1,005	1,910
Total	5,682	455	12,059	18,196

In this case the HVAC estimate was about 6,100 kWh annually compared to an actual HVAC energy use of 6,000 kWh. This is an example of a house where the estimate does a good job of estimating the actual energy use. While houses 1 and 3 have each had positive regression results, let's examine what happens when you get a negative estimate for an end use. The results for house 10 are shown in Table 8-11.

Table 8-11. Regression results for house 10.

Parameter	End Use	Coefficients	Standard Error	t Stat
a_1	Heating	0.60	0.33	1.82
a_2	Cooling	-1.65	1.98	-0.83
a_3	Other	290.8	30.5	9.5

The results for house 10 indicate that the cooling energy use is negative because it has a negative coefficient, which defies physical reality. Upon examining the t value for house 10, it once again appears that cooling is not a statistically significant parameter.

There are two common methods for dealing with this anomaly: the first method is to combine the statistically insignificant parameters with significant ones (e.g., we could combine the air conditioning and heating parameters and try to estimate the monthly HVAC use for this house). A second method is to simply discard the cooling parameter and perform the regression using only the heating and other parameters. Let's examine how each of these methods perform.

If we combine the heating and cooling parameters into a single HVAC parameter (by adding the cooling and heating degree days for each month), we generate the results presented in Table 8-12.

Table 8-12. HVAC regression results for house 10.

Parameter	End Use	Coefficients	Standard Error	t Stat
a_1	HVAC	0.89	0.25	3.59
a_3	Other	260.1	19.9	13.1

Using these regression coefficients, the following end-use estimates are produced for the house:

Month	Heating	Other	Total
January	722	1,561	2,283
February	544	1,561	2,105
March	528	1,561	2,089
April	391	1,561	1,952
May	240	1,561	1,801
June	144	1,561	1,704
July	137	1,561	1,698
August	154	1,561	1,715
September	137	1,561	1,697
October	296	1,561	1,857
November	522	1,561	2,083
December	667	1,561	2,227
Total	4,483	18,729	23,211

We estimate a total annual HVAC use of about 4,500 kWh compared to an actual HVAC use of 7,000 kWh. However, if we simply ignore the air conditioning end use we get the regression results shown in Table 8-13.

Table 8-13. Heating-only regression results for house 10.

Parameter	End Use	Coefficients	Standard Error	t Stat
a_1	Heating	0.81	0.21	3.90
a_3	Other	269.5	16.4	16.4

Using these regression results, we obtain the following end-use breakdown:

Month	Heating	Other	Total
January	657	1,617	2,274
February	495	1,617	2,112
March	480	1,617	2,097
April	355	1,617	1,972
May	213	1,617	1,830
June	96	1,617	1,713
July	28	1,617	1,645
August	41	1,617	1,658
September	90	1,617	1,707
October	269	1,617	1,886
November	475	1,617	2,092
December	606	1,617	2,223
Total	3,807	19,404	23,211

In this case, we estimate an annual heating usage of about 3,800 kWh, which is significantly lower than the estimate obtained by combining the parameters.

Hourly CDA Analysis

Now that we have examined annual and monthly CDA analysis let's take a look at hourly CDA analysis. While monthly and annual CDA analysis can be used to disaggregate billing history into end-use components, hourly CDA can be used to disaggregate hourly whole-building load research data into hourly end-use loads. However, in order to successfully generate end-use estimates you will need fairly stringent proxy estimators for each variable. This section will present some proxy estimators of the types that have been successfully used in the past for various end uses (Regional Economic Research, 1991).

Space Heating

Space heating is a function of several parameters including:

- the capacity of the heating system
- the temperature difference between the indoor space and outdoor space
- shell losses through the walls, roof, and windows
- the portion of the building which is heated

One specification of space heating would be:

$$Heat = h_1 \cdot HeatCap \cdot T_{diff} + h_2 \cdot Heatcap \cdot T_{diff} \cdot UA + h_3 \cdot HeatSq \cdot T_{diff} + h_4 \cdot HeatSq \cdot UA + h_5 \cdot HeatCap \cdot T_{diff} \cdot Occupants$$

(Equation 8-3)

where:

$Heat$ = the space heating estimate for the hour

$HeatCap$ = the heating capacity

T_{diff} = the temperature difference between the inside space and outdoor

UA = the overall building UA

$HeatSq$ = the heated square footage of the building

$Occupants$ = the percent of occupants in the building.

The overall building UA of the building can be calculated as shown in Equation 8-4. The U value is the inverse of the more commonly recognized R value (e.g., R-11 insulation) and provides a measure of how much heat can flow through the shell of a building.

$$UA = U_{wall} \cdot A_{wall} + U_{roof} \cdot + U_{window} \cdot A_{window}$$ **(Equation 8-4)**

where:

U_{wall} is the U value of the walls

A_{wall} is the area of the walls

U_{roof} is the U value of the roof

A_{roof} is the area of the roof

U_{window} is the U value of the window

A_{window} is the area of the window.

Air Conditioning

Air conditioning also contains many parameters, including:

- the capacity of the cooling system
- the temperature difference between the indoor space and outdoor space
- shell gains through the walls, roof, and windows
- solar gains through the windows
- the portion of the building which is cooled

The cooling specification is therefore similar to that of the heating specification:

$$Cool = c_1 \cdot CoolCap \cdot T_{diff} + c_2 \cdot CoolCap \cdot T_{diff} \cdot UA +$$
$$c_3 \cdot Area_{window} \cdot Gain_{solar} + c_4 \cdot CoolSq \cdot T_{diff} +$$
$$c_5 \cdot CoolSq \cdot UA + c_6 \cdot Occupants \quad \textbf{(Equation 8-5)}$$

where:

   $Cool$ = the air conditioning estimate for the hour

   $CoolCap$ = the cooling capacity

   T_{diff} = the temperature difference between the inside space and outdoor space

   UA = the overall building UA

   $Area_{window}$ = the area of the windows

   $Gain_{solar}$ = the solar gain through the windows

   $CoolSq$ = the cooled square footage of the building

   $Occupants$ = the percent of occupants in the building.

Ventilation

In residential buildings, the ventilation end use (supply fans, etc.) is generally included in the heating and cooling end uses since the ventilation energy use is not a significant contributor to the building load and the ventilation system only operates when the heating or cooling system operates. However, in commercial and industrial facilities, the ventilation end use can easily account for 5 to 10 percent of the building load at any one time. The factors which affect the ventilation end-use energy include:

- the capacity of the ventilation system
- the schedule of the ventilation system

A specification for the ventilation system might be:

$$Vent = v_1 \cdot VentCap + v_2 \cdot VentCap \cdot Schedule + v_3 \cdot VentCap \cdot Occupants + v_4 \cdot VentCap \cdot Schedule \cdot Occupants$$

(Equation 8-6)

where:

   $Vent$ – the ventilation system estimate for the hour

   $VentCap$ = the ventilation capacity (in kW)

Schedule = the fraction of ventilation currently based upon schedule data collected on site

Occupants = the percent of occupants in the building.

Indoor Lighting

Next to space heating and air conditioning, indoor lighting is probably the largest end use in a significant number of buildings. The control method used to turn the lights on or off can be a significant factor in determining when lights are being used. For example, while interior offices with little or no window area are predominantly controlled based on whether or not they are being used, exterior office space which has daylight dimmer controls on lights might be controlled by the amount of light entering the space. Another significant parameter may be the type of space being lit. While hallways and lobbies may have lights on during business hours, offices may have lights on during a portion of the business hours, and conference rooms may only have lights on during a small fraction of the business day. Indoor lighting use is a function of several parameters, including:

- the capacity of the lights
- the type of control system used to control the lights
- daylight entering the lighted area
- the occupancy of the building
- the time that lights are scheduled to be on
- the type of space being lit

An example of a lighting model specification would be:

$$Light = l_1 \cdot LightCap = l_2 \cdot LightCap \cdot Schedule + l_3 \cdot \frac{Daylight}{LightCap} +$$
$$l_4 \cdot LightCap \cdot SpaceType + l_5 \cdot LightCap \cdot Schedule \cdot SpaceType +$$
$$l_6 \cdot LightCap \cdot \frac{Daylight}{LightCap} \cdot SpaceType + l_7 \cdot LightCap \cdot Occupants +$$
$$l_8 \cdot LightCap \cdot Schedule \cdot Occupants + l_9 \cdot LightCap \cdot \frac{Daylight}{LightCap} \cdot Occupants$$

(Equation 8-7)

where:

LightCap = the capacity of the lighting system

Schedule = the scheduled fraction of lights which are on at any given time

Daylight = the amount of daylight entering the space, in lumens

SpaceType = a flag to indicate space type (0 for office space, 1 otherwise)

Occupants = the percent of occupants in the building.

Hot Water

Hot water usage differs greatly between the three major sectors: residential, commercial, and industrial. In residences, the peak hot water usage tends to be in the morning when people shower or in the evening when people run their dishwashers and do laundry. Fairly low hot water usage occurs during the middle of the day with the nighttime hours having the least usage. In commercial spaces, however, hot water tends to be used as more people occupy the space due to increased hand washing, etc. In industry, water is used for domestic purposes but a far greater amount of water may be used in industrial processes to either provide heat or cool processes.

The hot water model presented here is intended primarily for domestic hot water use and not industrial processes. The variables which affect the hot water usage in a facility include:

- the capacity of the hot water system (in kW or MBtu)
- the scheduled usage of the water
- the number of occupants in a building at any one time

One hot water model is:

$$Water = w_1 \cdot WaterCap + w_2 \cdot WaterCap \cdot Schedule + w_3 \cdot WaterCap \cdot Occupancy$$ **(Equation 8-8)**

where:

> *Water* = the energy use of the water heating system
> *WaterCap* = the capacity of the water heating system
> *Schedule* = the scheduled use of the hot water
> *Occupancy* = the percent of occupants in the building.

Cooking

In commercial cooking, the energy use is largely a function of how heavily the cooking appliances are being utilized, and hence is highly dependent upon the occupancy in the restaurant. The factors which affect cooking energy use are:

- the cooking capacity of the equipment
- the cooking schedule
- the occupancy of the restaurant

One form of a cooking end-use model is:

$$Cook = c_1 \cdot CookCap + c_2 \cdot CookCap \cdot Schedule + c_3 \cdot CookCap \cdot Occupancy$$

(Equation 8-9)

where:

> *Cook* = the energy use of the cooking appliances
> *CookCap* = the capacity of the cooking appliances
> *Schedule* = the scheduled use of the cooking appliances
> *Occupancy* = the percent of occupants in the building.

Refrigeration

In restaurants and grocery stores, the energy used for refrigeration/freezer equipment can be a significant portion of the energy bill. The factors which affect refrigeration energy use are:

- the capacity of the refrigeration compressors
- the location of the refrigeration compressors

- the temperature difference between inside and outside
- the occupancy of the building
- the type of refrigeration system

One form of a refrigeration end-use model is:

$$Refrig = r_1 \cdot CompCap + r_2 \cdot CompCap \cdot Location \\ + r_3 \cdot CompCap \cdot T \cdot Location + r_4 \cdot CompCap \cdot Occupancy \\ + r_5 \cdot CompCap \cdot Type$$

(Equation 8-10)

where:

$Refrig$ = refrigeration energy use

$CompCap$ = the capacity of the refrigeration compressors (in kW)

$Location$ = compressor location flag (0 for indoors, 1 for outdoors)

T = outside temperature

$Occupancy$ = number of people in the store

$Type$ = refrigerator/freezer type [0 for closed (e.g., door, walk-in freezer, etc.), 1 for open (e.g., coffin case, etc.)].

Exterior Lighting

The exterior lighting end use consists of parking lot lights, security lights, street lamps, and other light sources which may be connected to the meter but are outside of structures. The exterior lighting is largely a function of:

- the capacity of the lights
- the schedule of the lights
- the amount of sunlight present

An exterior lighting model is:

$$Ext = e_1 \cdot ExtCap + e_2 \cdot ExtCap \cdot Schedule \\ + e_3 \cdot ExtCap \cdot (1 - Schedule) \cdot Sunlight$$

(Equation 8-11)

where:

Ext = the hourly energy use for exterior lighting

$ExtCap$ = the exterior lighting capacity

$Schedule$ = the fraction of hour lights are scheduled on

$Sunlight$ = the fraction of hour that daylight is occurring.

Miscellaneous

The miscellaneous, or other, end use is really a catch-all for all of the energy use which has not been captured in the other areas. The miscellaneous end use is a function of:

- the capacity of the miscellaneous equipment
- the schedule of the equipment
- the number of occupants

The miscellaneous end use might have a form similar to:

$$Misc = m_1 \cdot MiscCap + m_2 \cdot MiscCap \cdot Schedule + m_3 \cdot MiscCap \cdot Occupancy$$

(Equation 8-12)

where:

$Misc$ = the miscellaneous energy use for the hour

$MiscCap$ = the capacity of the miscellaneous equipment

$Schedule$ = the miscellaneous equipment schedule

$Occupancy$ = the percent of occupancy in the building.

Hourly CDA Statistical Analysis

Once you have developed initial regression equations for all of your end uses, you can start performing the statistical analyses. The steps involved in the analysis are:

- determine how to segment your sample
- within each segment, you need to:
- run analysis using statistical models
- evaluate analysis to improve models
- run analysis using final statistical models

Historically, utilities have generally desired that results be generated by building type and hence the analysis was performed using this same segmentation scheme. The tendency for energy professionals is to assume that end-use energy among buildings has a similar pattern and magnitude within a given type of building (e.g., all retail stores will exhibit a similar behavior). While this is certainly true to a large extent, it may be that there are other factors which are stronger drivers such as the square footage of the building, the weather at the building location, etc. During the analysis phase of the project, try resegmenting the buildings to see if another segmentation may be more appropriate. If you remember one thing during your project it's that you shouldn't be afraid to try something new—you may discover something revolutionary.

The overall regression equation for the hourly CDA analysis is obtained by summing the respective end-use equations as shown in Equation 8-13.

$$Load = Heat + Cool + Vent + Light + Water + Cook + Refrig + Ext + Mis$$

(Equation 8-13)

where:

$Load$ = hourly total building load

$Heat$ through $Misc$ = end use (Equations 8-3 through 8-12).

Additionally, when implementing statistical models there are two methods of using the CDA equations. The first method is to absolute loads (kW, etc.) for each building and the second method is normalized loads (kW/ft^2, etc.) for the buildings. If you use the absolute loads, the buildings with the higher loads may tend to dominate or distort the statistical analysis. For example, if you have 30 buildings with an average load of 10 kW each and two building with an average load of 75 kW, the two highest buildings may drive the statistical analysis, which can lead to results that work quite well for the two buildings but are not representative of the other 28 buildings. By normalizing the loads and end-use proxy estimators you help to flatten out any variation in the magnitude of loads between buildings—leading to results which are more representative for all of the buildings in a segment.

Statistically Adjusted Engineering Models

SAE models, which are a subclass of CDA models, are used to quickly reconcile engineering estimates with a known total. In a SAE model, a known load is regressed against estimates, such as end-use estimates, and a coefficient is obtained for each end use (and hour if regression is performed on an hourly basis). Next, each end use is multiplied by the reciprocal of the associated coefficient. The basic form of the SAE regression model is shown in Equation 8-14.

$$Load = a_1 \cdot end\text{-}use_1 + a_2 \cdot end\text{-}use_2 + \cdots + a_n \cdot end\text{-}use_n$$

(Equation 8-14)

where:

$Load$ = the known total load at a given hour

$End\ use$ = the end-use load estimates for the hour.

After the coefficients are obtained for each hour, the estimates are multiplied by the reciprocal of the coefficient to obtain the statistically adjusted estimate as shown in Equation 8-15.

$$end\ use_{n,adj} = \frac{1}{a_n} \cdot end\ use_n \qquad \textbf{(Equation 8-15)}$$

where:

$End\ use_{n,adj}$ = the adjusted estimate for end use n

$End\ use_n$ = the original estimate for end use n.

Example 8-4: SAE Analysis

In this example, we have a group of 20 office buildings which have a varied mix of end uses connected to load research meters. Each building can have up to five end uses: air conditioning, lighting, hot water, miscellaneous, and exterior lighting. The end-use mix for each of the 20 building is presented in Table 8-14. The hourly end-use loads and load research data for each of the buildings is shown in the Appendix in Table A-1.

Table 8-14. End uses connected to load research meter.

Site	Air Conditioning	Lights	Hot Water	Miscellaneous	Exterior Lighting
1		X		X	X
2	X	X	X	X	X
3	X	X		X	X
4		X		X	
5	X	X	X		X
6	X	X		X	X
7		X		X	
8	X	X	X	X	X
9	X	X		X	X
10		X		X	X
11	X	X	X	X	X
12	X	X		X	X
13		X		X	X
14	X	X	X	X	X
15	X	X		X	X
16		X		X	X
17	X	X	X		X
18	X	X		X	X
19		X		X	X
20	X				

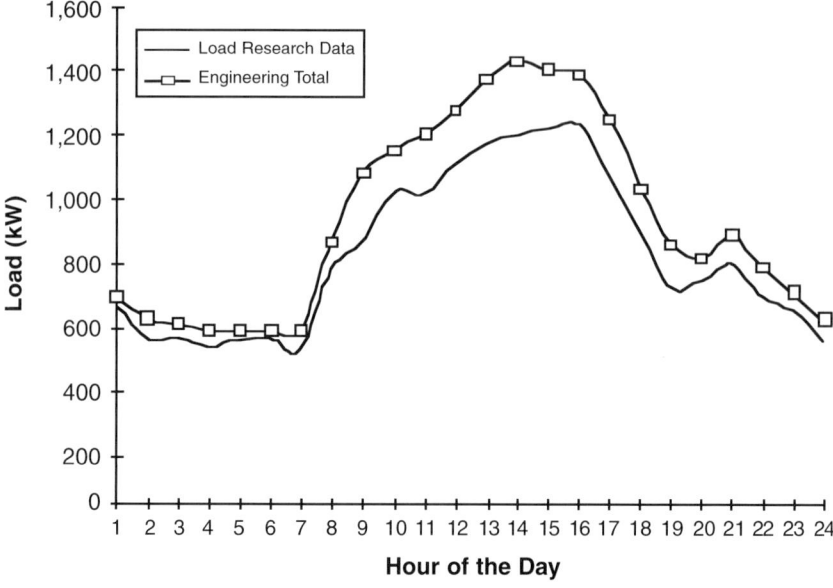

Figure 8-1. Comparison of engineering total and load research total.

The total end use is calculated by summing the end uses for each hour and building. A visual method which can be used to determine how well the engineering totals match the load research data can be seen in Figure 8-1, which presents the sum of the engineering and load research end use for all of the buildings.

After the regression analysis was performed, the coefficients shown in Table 8-15 were generated for the end uses. Since the regression analysis was performed for all of the buildings and all hours as a single group, only one coefficient was generated for each end use. If we apply these coefficients to every hourly end use for each building and sum the adjusted end uses, we obtain an adjusted engineering total for each building and hour.

Table 8-15. Regression coefficients.

End Use	Coefficients	Standard Error	t Stat
Air conditioning	0.857	0.017	51.56
Indoor lighting	0.924	0.031	29.56
Hot water	0.776	0.269	2.89
Miscellaneous	0.757	0.050	14.99
Exterior lighting	0.922	0.016	58.82

Although we know that on average the adjusted engineering values will match the load research data, how well do these compare on an hourly basis for all of the buildings? Figure 8-2 presents a comparison of the adjusted engineering total to the load research data. If you wanted to match the adjusted engineering total to the load research for every hour, you would have to perform a separate regression analysis for each hour across all of the buildings.

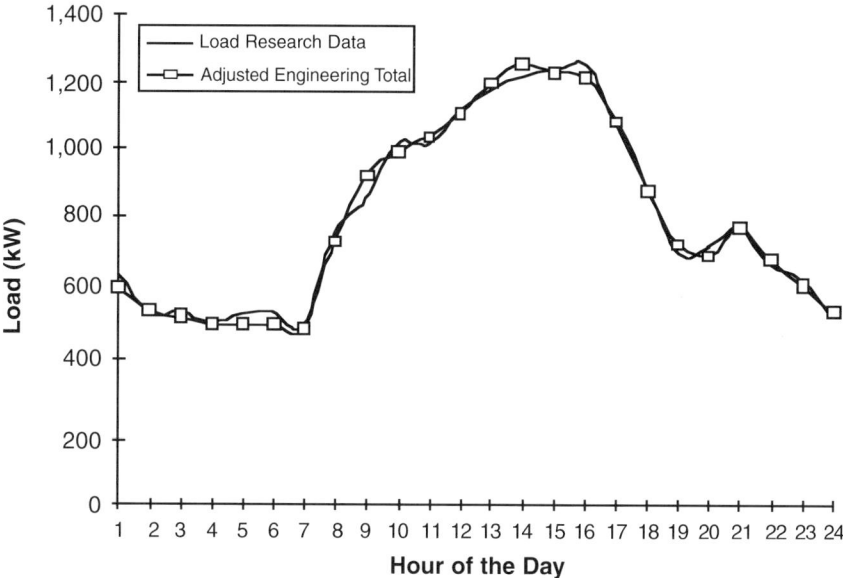

Figure 8-2. Comparison of adjusted engineering total and load research total.

Hybrid Statistical Engineering Method

To the best of my knowledge, the HSEM was developed by XENERGY, Inc. in the late 1980s in an effort to combine the best of engineering analysis and statistical analysis. The HSEM process starts the same as the SAE model; a known load is regressed against estimates, such as end-use estimates, and a coefficient is obtained for each end use (and hour if regression is performed on an hourly basis). However, where the SAE model applies the reciprocal of the coefficient to determine an adjusted end-use value, the HSEM method simply uses the results of the regression analysis as an indicator of how well the engineering model is performing as well as which end uses may be under- or over-predicting on average.

If a regression coefficient is greater than unity it implies that the end use estimated is understated and conversely if a regression coefficient is determined to be less than unity the implication is that the end use estimate is overstated. When using regression coefficients in this manner, it's important to pay attention to the t statistic associated with the regression coefficient and only use those results which are statistically significant.

Furthermore, the HSEM analysis is generally performed for two sets of hours (business and non-business) as well as different seasons to determine how well the engineering HVAC model is performing throughout the year.

With HSEM analysis, it is common to find three systematic errors with the end-use data collected in the field: first, end uses (especially lighting and miscellaneous) tend to be overstated during the business hours; second, these same end uses tend to be understated during the non-business hours of a building; and third, people may have a misconception about when their building is occupied—it's common that people are working in a building before and after official business hours. All three of these phenomenon are shown in Figure 8-3.

The overstatement of loads during the day can be largely attributed to the following:

- people stating that 100 percent of their lights and miscellaneous equipment are on during the day
- the misconception that equipment operates at full (or nameplate) load

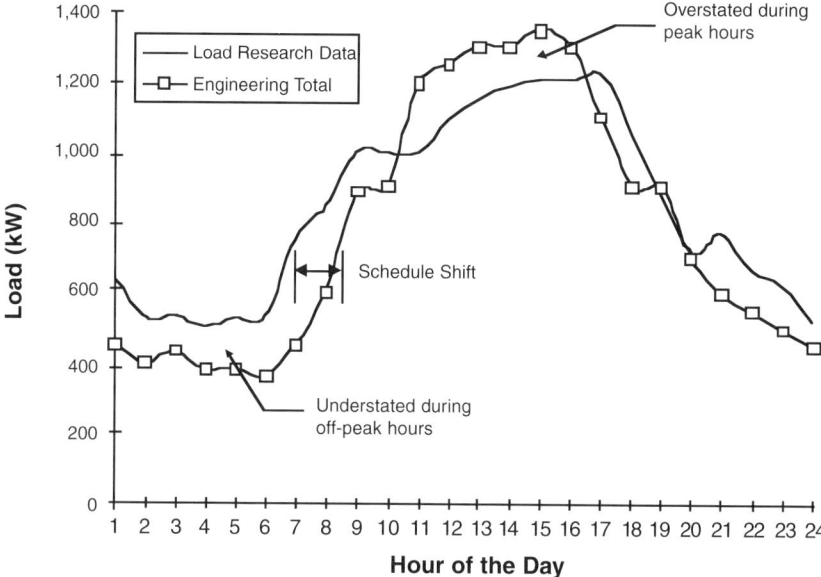

Figure 8-3. Common HSEM findings.

Conversely, the understatement of loads during the nighttime hours is due to:

- people believing that none of the lights in their facility are on during non-business hours
- people stating that their miscellaneous equipment is turned off during non-business hours

Load Shape Aggregation

Whichever statistical method you have used, remember to aggregate your results to the desired population. The aggregation techniques are the same as those presented in Chapter 6.

References

Marion, William, and Urban, Ken, "User's Manual for TMY2s," National Renewable Energy Laboratory, June 1996.

"New England Power Service Company: Commercial End Use Load Shape Analysis," Regional Economic Research, Inc., July 1991.

9
Transferring Load Shapes

You Want to What?

There are two major reasons that utilities wish to transfer load shapes: the first is to transfer load shapes from one region to another (within a service territory or among different regions of the country); the second reason is to transfer load shapes over time. This chapter will discuss some simple methods of how to perform both of these tasks.

In the case of transferring loads between utilities, it is customary to designate the original utility as the donor utility and the receiving utility as the recipient. The three primary techniques which will be examined in this chapter include the class buildup, class adjustment, and hybrid techniques—all of which can be used to transfer either whole-building load shapes or end-use/appliance load shapes (Linder, K.P., and Breese, J.S., July 1984).

Major factors which affect how readily load shapes can be transferred from one location to another include differences in weather climates, end-use/appliance saturation, and end-use efficiency. When transferring load shapes over time the concerns include economic factors (such as the price of energy), changes in efficiency of equipment, the introduction rate of new equipment in the service territory, and the removal of old equipment.

The Future of Load Shape Transfers

It is my belief that load shape transfer between utilities will cease to exist as a primary form of developing end-use load shapes in a deregulated utility environment. As deregulation comes of age, utilities will be less likely to share competitive information (such as market share and appliance saturation) with other utilities—and without this information, transferring load shapes becomes a difficult task yielding large uncertainties in the resulting load shapes. This trend may be offset, however, by the increased use of load-shape transfer within utilities, especially as companies merge, encompass larger geographical areas, and have a greater need to share data internally.

Class Load Buildup

If your goal is to transfer only the whole-building loads between utilities then the class load buildup technique is one of the simplest methods which may be used. The basic buildup procedure is to take stratified loads which have been developed for a customer sector or class at Utility A and transfer them to Utility B via the use of expansion factors as shown in Equation 9-1.

$$R_{Lh} = \sum_{s} D_{Lh,s} \cdot EF_s \qquad \text{(Equation 9-1)}$$

where:

R_{Lh} = the recipient end use or class/sector load at hour h

$D_{Lh,s}$ = the donor end use or class/sector load for stratum s at hour h

EF_s = the expansion factor for stratum s.

Example 9-1: Whole-Building Buildup Example

The average customer hourly, whole-building load data for typical grocery stores in each of four stratum for a generic utility single daytype are shown in Table 9-1.

Table 9-1. Donor utility average grocery hourly loads.

Hour	Stratum 1	Stratum 2	Stratum 3	Stratum 4
1	12.1	42.1	114.7	223.4
2	13.2	38.9	120.6	230.2
3	12.9	40.3	135.4	218.4
4	12.5	41.2	128.7	234.7
5	15.8	52.3	146.6	255.8
6	14.9	64.5	177.2	268.5
7	22.4	70.5	197.6	290.6
8	31.2	89.5	192.1	288.1
9	39.4	94.6	195.0	289.7
10	38.9	93.4	188.6	285.8
11	36.0	84.9	166.4	257.3
12	34.3	88.3	174.9	257.2
13	35.7	88.3	174.7	264.8
14	34.6	89.2	176.6	270.3
15	33.6	84.9	175.1	265.3
16	35.5	88.7	170.5	258.4
17	34.2	84.6	176.0	256.3
18	34.6	92.8	174.8	270.3
19	30.1	92.5	188.9	289.2
20	19.8	82.3	197.3	296.0
21	18.2	69.7	187.9	281.8
22	15.7	53.7	184.1	268.1
23	16.9	44.8	136.4	256.9
24	17.0	43.2	126.9	218.6

Given the expansion factors for each of the four stratum for the recipient utility as shown below, construct the grocery sector whole-building load for the recipient utility. Note that the expansion factors used in this case are the number of grocery customers in each strata for the recipient utility.

Recipient Stratum	Number of Customers
1	40
2	12
3	2
4	1

Step 1: Apply the expansion factors to each hour of every stratum of the donor load data using Equation 9-1. The results of this manipulation are shown in Table 9-2.

Step 2: Sum the hour data across the strata to obtain the grocery sector whole-building load estimate, also shown in Table 9-2.

Table 9-2. Recipient utility hourly stratum and sector loads.

	Stratum				
1	2	3	4		Sector Total
484	505	229	223		1,442
528	467	241	230		1,466
516	484	271	218		1,489
500	494	257	235		1,487
632	628	293	256		1,809
596	774	354	269		1,993
896	846	395	291		2,428
1,248	1,074	384	288		2,995
1,576	1,135	390	290		3,391
1,555	1,120	377	286		3,338
1,441	1,019	333	257		3,049
1,372	1,060	350	257		3,038
1,426	1,059	349	265		3,099
1,384	1,070	353	270		3,078
1,344	1,019	350	265		2,978
1,419	1,065	341	258		3,083
1,367	1,015	352	256		2,990
1,384	1,114	350	270		3,118
1,204	1,111	378	289		2,982
792	988	395	296		2,470
728	836	376	282		2,222
628	644	368	268		1,909
676	538	273	257		1,743
680	518	254	219		1,671

Example 9-2: End-Use Buildup Example

An office sector from a donor utility has stratum-level hourly air-conditioning loads as given in Table 9-3.

Table 9-3. Donor utility stratum air-conditioning loads.

Hour	Stratum 1	Stratum 2	Stratum 3	Stratum 4
1	300	1,432	1,370	122
2	313	1,279	1,028	100
3	346	1,146	901	112
4	324	1,087	915	102
5	299	1,002	877	107
6	309	1,029	946	110
7	601	1,883	1,256	112
8	956	1,937	1,292	115
9	1,321	2,025	1,350	120
10	1,798	2,133	1,422	126
11	1,845	2,251	1,501	133
12	1,957	2,366	1,577	140
13	2,172	2,444	1,629	145
14	2,250	2,531	1,688	150
15	2,304	2,592	1,728	154
16	2,568	2,622	1,748	155
17	2,355	2,649	1,766	157
18	1,856	2,639	1,760	156
19	1,432	2,294	1,748	155
20	1,107	1,985	1,694	151
21	803	1,654	1,638	146
22	700	1,436	1,602	142
23	713	1,439	1,550	138
24	688	1,399	1,375	122

The number of customers per stratum are shown on the following page are for both the donor and recipient utility. Ignoring weather effects, determine the office sector air-conditioning load for the recipient utility.

Stratum	Donor Customers	Recipient Customers
1	265	187
2	146	68
3	67	19
4	3	1

Step 1: Determine the average hourly loads for the offices in each stratum of the donor utility by dividing the total stratum load by the number of donor utility customers, as shown in Table 9-4.

Table 9-4. Donor utility average stratum air-conditioning loads.

	Stratum			
Hour	1	2	3	4
1	1.1	9.8	20.5	40.6
2	1.2	8.8	15.3	33.4
3	1.3	7.8	13.4	37.3
4	1.2	7.4	13.6	34.1
5	1.1	6.9	13.1	35.5
6	1.2	7.1	14.1	36.7
7	2.3	12.9	18.7	37.2
8	3.6	13.3	19.3	38.3
9	5.0	13.9	20.1	40.0
10	6.8	14.6	21.2	42.1
11	7.0	15.4	22.4	44.5
12	7.4	16.2	23.5	46.7
13	8.2	16.7	24.3	48.3
14	8.5	17.3	25.2	50.0
15	8.7	17.8	25.8	51.2
16	9.7	18.0	26.1	51.8
17	8.9	18.1	26.4	52.3
18	7.0	18.1	26.3	52.1
19	5.4	15.7	26.1	51.8
20	4.2	13.6	25.3	50.2
21	3.0	11.3	24.4	48.5
22	2.6	9.8	23.9	47.5
23	2.7	9.9	23.1	45.9
24	2.6	9.6	20.5	40.7

Step 2: Apply the expansion factors to each hour of every stratum of the average donor load data using Equation 9-1. The results of this manipulation are shown in Table 9-5.

Table 9-5. Recipient utility stratum and sector air-conditioning loads.

Hour	Stratum 1	Stratum 2	Stratum 3	Stratum 4	Sector Total
1	212	667	389	41	1,308
2	221	596	292	33	1,141
3	244	534	255	37	1,070
4	228	506	259	34	1,028
5	211	467	249	36	962
6	218	479	268	37	1,002
7	424	877	356	37	1,694
8	675	902	366	38	1,982
9	932	943	383	40	2,298
10	1,269	993	403	42	2,708
11	1,302	1,048	426	44	2,821
12	1,381	1,102	447	47	2,977
13	1,533	1,138	462	48	3,181
14	1,588	1,179	479	50	3,295
15	1,626	1,207	490	51	3,374
16	1,812	1,221	496	52	3,581
17	1,662	1,234	501	52	3,449
18	1,310	1,229	499	52	3,090
19	1,011	1,069	496	52	2,627
20	781	925	480	50	2,236
21	567	770	465	49	1,850
22	494	669	454	47	1,665
23	503	670	440	46	1,658
24	485	651	390	41	1,567

Step 3: Sum the hour data across the strata to obtain the grocery sector whole-building load estimate as presented in Table 9-5.

Class Load Adjustment

In the previous section, we saw how the class load buildup technique was simple and quick to implement. However, its success is largely dependent upon how similar the donor and recipient utility are in terms of both load shape patterns and end-use market shares. The class load adjustment method provides an advantage over the basic buildup method by taking some of the differences between donor and recipient utilities into account when load shapes are being transferred among utilities or regions.

The class load process involves:

- obtaining customer average end-use load shapes from the donor utility for the major end uses
- obtaining appliance saturation for both the donor and recipient utilities
- aggregating the end-use shapes to the donor class level by the number of customers and appliance/end-use saturation
- subtracting the known end uses from the class total load to obtain a residual donor load shape
- calculating the residual recipient load shape
- for non-weather-sensitive loads, obtaining a residual recipient end-use shape by multiplying the residual end-use shape by the ratio of the number of recipient customers over the number of donor customers
- for weather-sensitive loads, adjusting the end uses for weather and system efficiency
- the total recipient class load is the sum of the class-level recipient end-use loads plus the recipient residual load

The basic equation for aggregating end-use loads to class level is shown as Equation 9-2.

$$L_{c,eh} = L_{eh} \cdot A_e \cdot Ef_c \qquad \textbf{(Equation 9-2)}$$

where:

$L_{c,eh}$ = the class-level load for end use e at hour h
L_{eh} = the average customer load for end use e at hour h
A_e = the appliance/end-use saturation for end use e
Ef_c = the expansion factor for the class.

The residual load for the donor utility may be calculated using Equation 9-3.

$$R_{dh} = L_{c,dh} - \sum_e L_{c,eh} \qquad \text{(Equation 9-3)}$$

where:

R_{dh} = the residual donor load at hour h
$L_{c,dh}$ = the total class load for the donor utility at hour h
$L_{c,eh}$ = the class-level end-use loads for the donor utility at hour h.

The adjusted recipient residual load may be calculated using Equation 9-4.

$$R_{rh} = R_{dh} \cdot \frac{N_r}{N_d} \qquad \text{(Equation 9-4)}$$

where:

R_{rh} = the recipient residual load at hour h
R_{dh} = the donor residual load at hour h
N_r = the number of recipient customers
N_d = the number of donor customers.

Air Conditioning and Space-Heating

For weather-sensitive loads, which include air conditioning and space heating, you may also need to adjust for differing weather conditions between the donor and recipient utility. For air-conditioning cooking degree days can provide a reasonable adjustment factor; similarly, heating degree days provide a readily available adjustment factor for space heating. Additionally, you can adjust both of these end uses for differences in system efficiency between the donor and recipient utilities.

Equation 9-5 presents a means of adjusting the air-conditioning loads between regions as a function of cooling degree days and cooling efficiency. Equation 9-6 shows a similar representation for adjusting space heating end uses.

$$AC_{rh} = AC_{dh} \cdot A_r \cdot Ef_r \cdot \frac{CDD_r}{CDD_d} \cdot \frac{EER_d}{EER_r} \quad \text{(Equation 9-5)}$$

where:

AC_{rh} = recipient air-conditioning load at hour h

AC_{dh} = donor air-conditioning load at hour h

A_r = recipient appliance/end-use saturation

Ef_r = recipient expansion factor

EER_r = recipient air-conditioning energy efficiency ratio (EER)

EER_d = donor air-conditioning energy efficiency ratio.

$$SH_{rh} = SH_{dh} \cdot \frac{HDD_r}{HDD_d} \cdot \frac{\eta_r}{\eta_d} \quad \text{(Equation 9-6)}$$

where:

SH_{rh} = recipient space heating load at hour h

SH_{dh} = donor space heating load at hour h

η_r = recipient heating efficiency

η_d = donor heating efficiency.

Example 9-3: Restaurant Transfer Example
You are provided with the following data for a restaurant on a peak May weekday:
- donor average customer loads (Table 9-6)
- donor total class load (Table 9-6)

Table 9-6. Donor utility end-use loads for May peak weekday.

Hour	Air Conditioning	Interior Light	Hot Water	Cooking	Refrigeration	Ventilation	Exterior Light	Class Total
1	5.3	1.4	0.2	0.2	2.6	1.9	0.6	5,311
2	2.5	0.8	0.2	0.2	2.5	1.0	0.3	3,252
3	1.5	0.9	0.2	0.2	2.5	1.0	0.3	2,951
4	2.0	0.9	0.2	0.2	2.4	1.0	0.3	3,092
5	1.8	0.9	0.2	0.2	2.4	1.0	0.3	3,413
6	1.7	0.9	0.4	0.2	2.4	1.0	0.3	2,934
7	2.1	2.1	0.7	1.0	2.5	1.0	0.0	4,482
8	2.0	2.2	0.7	1.1	2.5	1.0	0.1	4,611
9	2.7	2.6	0.9	1.7	2.6	1.0	0.1	5,613
10	5.6	2.7	0.9	1.8	2.7	2.0	0.1	7,187
11	6.3	3.3	0.9	2.3	2.9	2.0	0.1	8,983
12	7.5	3.5	1.1	2.6	3.2	2.1	0.0	9,694
13	8.3	3.2	1.5	2.7	3.2	2.1	0.0	9,565
14	8.2	3.2	2.0	2.7	3.1	2.1	0.0	9,670
15	8.2	3.2	1.5	2.6	2.9	2.1	0.0	9,788

(cont'd)

Table 9-6. Donor utility end-use loads for May peak weekday (cont'd).

Hour	Air Conditioning	Interior Light	Hot Water	Cooking	Refrigeration	Ventilation	Exterior Light	Class Total
16	8.2	3.2	1.1	2.5	3.0	2.1	0.0	9,367
17	8.2	3.2	0.9	2.2	3.2	2.1	0.4	8,618
18	8.7	3.3	1.1	2.2	3.5	2.1	0.5	10,801
19	9.2	3.4	2.2	2.6	3.4	2.1	0.6	11,268
20	9.0	3.5	2.2	2.6	3.4	2.1	0.9	12,086
21	8.3	3.0	1.7	2.4	3.0	2.1	1.0	10,725
22	7.6	3.0	1.1	2.2	3.0	2.1	1.0	9,242
23	7.4	2.8	0.7	1.8	2.8	2.1	0.8	8,071
24	6.0	2.0	0.2	0.4	2.6	1.9	0.7	6,441

- donor and recipient restaurant customers (Table 9-7)
- donor and recipient cooling degree days for the peak weekday in May (Table 9-7).
- donor and recipient air-conditioning EER (Table 9-7)

Determine the recipient total class load for the restaurants using the adjusted transfer method.

Table 9-7. Donor and recipient utility factors.

Factor	Donor Utility	Recipient Utility
Customers	400	250
CDD	10	7
EER	12	11.4

Step 1: Determine the average customer loads (see Table 9-6).

Step 2: Determine appliance saturation for donor and recipient utilities. Since we are dealing with a commercial facility (vs. a residential home) the end uses are a mix of appliance types and appliance quantities. Since we don't have better data, we will assume that the appliance saturation is 1 for both the donor and recipient utilities for all end uses. However, if you had an accurate idea of how the end-use saturations differ between the utilities you would apply that knowledge here.

Step 3: Aggregate the major donor end uses to the class level using Equation 9-2 where the expansion factor is the number of customers in the donor utility. The class-level end-use loads and estimated class-level total are shown in Table 9-8.

214 Chapter 9

Table 9-8. Donor class-level end-use loads.

Hour	Air Conditioning	Interior Light	Hot Water	Miscellaneous Equipment	Cooking	Refrigeration	Ventilation	Exterior Light
1	2,120	560	80	120	80	1,040	760	240
2	1,000	320	80	120	80	1,000	400	120
3	600	360	80	120	80	1,000	400	120
4	800	360	80	120	80	960	400	120
5	720	360	80	120	80	960	400	120
6	680	360	160	120	80	960	400	120
7	840	840	280	200	400	1,000	400	—
8	800	880	280	200	440	1,000	400	40
9	1,080	1,040	360	320	680	1,040	400	40
10	2,240	1,080	360	320	720	1,080	800	40
11	2,520	1,320	360	320	920	1,160	800	40
12	3,000	1,400	440	520	1,040	1,280	840	—
13	3,320	1,280	600	520	1,080	1,280	840	—
14	3,280	1,280	800	520	1,080	1,240	840	—

Hour	Air Conditioning	Interior Light	Hot Water	Miscellaneous Equipment	Cooking	Refrigeration	Ventilation	Exterior Light
15	3,280	1,280	600	520	1,040	1,160	840	—
16	3,280	1,280	440	520	1,000	1,200	840	—
17	3,280	1,280	360	520	880	1,280	840	160
18	3,480	1,320	440	560	880	1,400	840	200
19	3,680	1,360	880	560	1,040	1,360	840	240
20	3,600	1,400	880	560	1,040	1,360	840	360
21	3,320	1,200	680	360	960	1,200	840	400
22	3,040	1,200	440	360	880	1,200	840	400
23	2,960	1,120	280	240	720	1,120	840	320
24	2,400	800	80	200	160	1,040	760	280

Step 4: Determine the donor residual load from Equation 9-3, as shown in Table 9-9.

Step 5: Calculate the recipient residual load from Equation 9-4, as shown in Table 9-9.

Step 6: Calculate the recipient non-weather-sensitive loads using Equation 9-2. The results are shown in Table 9-10.

Table 9-9. Donor and recipient residual loads.

Residual Donor	Residual Recipient
311	195
132	82
191	119
172	108
573	358
54	34
522	326
571	357
653	408
547	342
1,543	964
1,174	734
645	403
630	394
1,068	668
807	504
18	11
1,681	1,051
1,308	817
2,046	1,279
1,765	1,103
882	551
471	295
721	451

Table 9-10. Recipient utility class end-use and total loads.

Air Conditioning	Interior Light	Hot Water	Miscellaneous Equipment	Cooking	Refrigeration	Ventilation	Exterior Light	Class Total
976	350	50	75	50	650	475	150	2,776
461	200	50	75	50	625	250	75	1,786
276	225	50	75	50	625	250	75	1,626
368	225	50	75	50	600	250	75	1,693
332	225	50	75	50	600	250	75	1,657
313	225	100	75	50	600	250	75	1,688
387	525	175	125	250	625	250	—	2,337
368	550	175	125	275	625	250	25	2,393
497	650	225	200	425	650	250	25	2,922
1,032	675	225	200	450	675	500	25	3,782
1,161	825	225	200	575	725	500	25	4,236
1,382	875	275	325	650	800	525	—	4,832
1,529	800	375	325	675	800	525	—	5,029
1,511	800	500	325	675	775	525	—	5,111
1,511	800	375	325	650	725	525	—	4,911

Air Conditioning	Interior Light	Hot Water	Miscellaneous Equipment	Cooking	Refrigeration	Ventilation	Exterior Light	Class Total
1,511	800	275	325	625	750	525	—	4,811
1,511	800	225	325	550	800	525	100	4,836
1,603	825	275	350	550	875	525	125	5,128
1,695	850	550	350	650	850	525	150	5,620
1,658	875	550	350	650	850	525	225	5,683
1,529	750	425	225	600	750	525	250	5,054
1,400	750	275	225	550	750	525	250	4,725
1,363	700	175	150	450	700	525	200	4,263
1,105	500	50	125	100	650	475	175	3,180

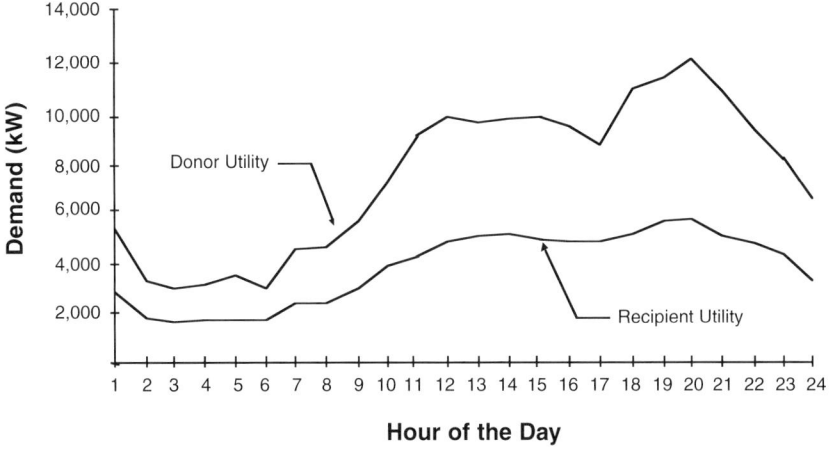

Figure 9-1. Donor and recipient total class loads.

Step 7: Calculate the recipient weather-sensitive (e.g. air-conditioning) loads using Equation 9-5 (see Table 9-10). In addition, we could have chosen to adjust the ventilation load depending on whether or not it varies with the air-conditioning load.

Step 8: Aggregate the recipient class-level and end-use and residual load to obtain the total recipient class load, as presented in Table 9-10.

The total class loads for both the donor and recipient utility are shown in Figure 9-1.

Class Load Hybrid

In the class load hybrid method, we integrate the buildup and adjustment techniques into an improved method. For each stratum in the class, we follow the steps in the adjustment method to develop end-use load shapes for each stratum. After data has been developed for all the stratum, we aggregate the results to form the class total end-use loads.

Reference

Linder, K.P., and Breese, J.S., "Load Data Transferability," EPRI EA-3255, Electric Power Research Institute, July 1984.

10
Beyond Load Shapes

A Means to an End

Often, load shapes in themselves are not the final solution that the utility is looking for, but rather a means to an end—perhaps a utility's goal is to obtain better data or additional data about its customers. But exactly which data can be derived or developed from load shapes may not be entirely clear. The purpose of this chapter is to provide simple techniques for fully utilizing load shapes to help ensure they are used to their full potential.

The data analyses which will be demonstrated in this chapter include how to calculate floorspace estimates, annual energy use, energy use indices, market shares, demand intensities, coincidence factors, and full-load hours.

Floorspace Estimates

The floorspace estimate (e.g., the square footage of the buildings) for a population can be developed by summing the products of the actual floorspace and population weight for all of the customers in a class, as presented in Equation 10-1. When you include a building in a load shape sample, the square footage of the buildings is one of the primary data elements which should be collected.

$$Floorspace_p = \sum_i \left(Floor_i \cdot W_i \right)$$ (Equation 10-1)

where:

$Floorspace_p$ = population floorspace estimate
$Floorspace_i$ = floorspace for customer i
W_i = population weight for customer i.

Example 10-1: Floorspace Estimation

Given the following floorspace and strata for the following buildings along with the population weight within each strata, develop an estimate of the population floorspace within each strata and for the total population.

Strata	Weight
1	30
2	25
3	7
4	3

Building	Strata	Annual Energy (kWh)	Square Footage
1	1	63,220	5,800
2	1	115,560	10,700
3	1	132,090	11,900
4	1	95,550	9,100
5	1	73,440	6,800
6	1	64,890	6,300
7	2	104,500	11,000
8	2	264,000	24,000
9	2	217,120	18,400
10	2	210,600	23,400
11	3	345,420	34,200
12	3	315,840	33,600
13	3	349,440	31,200
14	4	299,520	28,800
15	4	556,920	46,800

Step 1: Sum the product of the square footage and population weight for the buildings in each strata using Equation 10-1. The results are shown in Table 10-1.

Table 10-1. Strata and population floorspace estimates.

Strata	Floorspace Estimate (1,000s square foot)
1	1,518
2	1,920
3	693
4	227
Population	4,358

Step 2: Sum the strata floorspace estimates to obtain the population floorspace estimates. See Table 10-1 for results.

Monthly/Annual Energy Use

The monthly population energy use can be estimated by summing the products of the monthly energy use for each building by the population weight as shown in Equation 10-2. This equation assumes that the monthly billing history is available for each customer and month, or the monthly energy use of each end use has been determined. This approach works well for either end-use or whole-building energy consumption.

$$kWh_m = \sum_i (kWh_{mi} \cdot W_i) \qquad \text{(Equation 10-2)}$$

where:

kWh_m = the estimated population energy use for month m

kWh_{mi} = the monthly energy use for building i

W_i = the population weight for building i.

If the monthly billing history is unavailable, it can be obtained from whole-building load research data, if available, or it can be estimated by summing the product of the whole-building energy for each daytype in the month by the number of days each daytype represents as shown in Equation 10-3. This assumes that the daytypes are defined on either a monthly or smaller time period.

$$kWh_{mi} = \sum_{d}(kWh_d \cdot Days_d) \qquad \textbf{(Equation 10-3)}$$

where:

kWh_{mi} = the monthly energy use for building i

kWh_d = the daily energy use for daytype d

$Days_d$ = the number of days daytype d represents.

Example 10-2: Population Energy Estimation

Estimate the population annual energy use for the buildings presented in Example 10-1.

Step 1: Use Equation 10-2 to sum the weighted annual energy use from each building. The answer is 45,892 MWh—this was an easy example!

Energy Use Indices

Energy Use Indices (EUIs) provide a good means of comparing energy use among buildings since they normalize the annual energy use by dividing by the building square footage. To estimate an EUI for a group of buildings divide the population estimate of energy use by the population estimate of square footage for the buildings in question, as presented in Equation 10-4. This can be applied to either end-use or whole-building energy use.

$$EUI_{ef} = \frac{\sum_{i}(MBTU_{ief} \cdot W_i)}{\sum_{i}(Floor_{ief} \cdot W_i)} \qquad \textbf{(Equation 10-4)}$$

where:

EUI_{ef} = annual energy use indices for end use e and fuel f

kWh_{ief} = the annual energy use for building i and end use e and fuel f

$Floor_{ief}$ = the square footage of building i used for end use e and fuel f

W_i = the population weight for building i.

Example 10-3: Population EUI Estimates

Estimate the strata and the population end use EUIs for a group of buildings which have the annual end-use consumption shown in Table 10-2a. The population weights for each strata are also provided. One kWh equals 3.413 MBtu.

Strata	Population Weight
1	49.7
2	20.8
3	7

Step 1: Calculate the strata level end use EUIs using Equation 10-4. The results are shown in Table 10-2b.

Step 2: Calculate the population end use EUIs using Equation 10.4. Note that you cannot simply sum the strata EUIs as you are dealing with ratios instead of absolute numbers.

Market Shares

Market shares provide a measure of how much of your population square footage has a specific fuel and end-use combination. We are defining market share as a fraction of the population floorspace which uses the particular fuel/end-use mixture as shown in Equation 10-5.

$$MS_{ef} = \frac{\sum_i \left(Floor_{ief} \cdot W_i\right)}{\sum_i \left(Floor_i \cdot W_i\right)}$$ (Equation 10-5)

Table 10-2a. Annual Consumption Data for Example 10-3.

					Energy Consumption (kWh)				
Building	Strata	Square Footage	Air Conditioning	Space Heat	Interior Lighting	Hot Water	Miscellaneous Equipment	Ventilation	Exterior Lights
1	1	10,438	15,879	17,185	25,559	978	32,318	20,338	3,789
2	1	10,090	18,803	14,811	30,814	1,418	38,307	28,883	5,756
3	1	12,055	19,485	18,173	23,216	1,144	33,325	26,099	7,002
4	1	12,152	19,873	28,926	35,773	1,723	47,927	22,508	5,798
5	1	7,076	10,092	14,457	15,185	1,003	27,907	17,362	2,789
6	1	11,449	17,199	24,981	30,554	1,485	30,384	31,672	4,513
7	1	13,069	27,337	19,702	36,316	1,633	39,500	37,096	5,558
8	2	13,422	30,372	25,825	28,065	1,741	53,430	37,453	7,170
9	2	26,046	41,603	42,365	63,776	3,598	69,122	74,558	15,669
10	2	18,390	30,773	26,264	54,137	1,767	67,106	44,680	11,254
11	2	19,758	37,195	30,570	44,574	2,017	58,261	46,102	9,017
12	2	29,538	49,428	42,186	84,517	3,264	92,543	68,922	13,327
13	3	27,345	50,958	50,444	86,518	3,449	104,823	67,094	12,054
14	3	25,461	39,701	59,595	65,146	3,181	95,724	44,098	14,525
15	3	29,454	42,566	56,088	63,207	3,291	110,736	83,605	18,177

Table 10-2b. Strata and population end-use EUI estimates.

	EUI's (MBtu/Square Foot)						
Strata	Air Conditioning	Space Heat	Interior Lighting	Hot Water	Miscellaneous Equipment	Ventilation	Exterior Lights
1	5.75	6.18	8.83	0.42	11.16	8.23	1.57
2	6.03	5.33	8.76	0.39	10.84	8.65	1.80
3	5.53	6.89	8.92	0.41	12.92	8.08	1.86
Population	5.83	5.95	8.81	0.41	11.21	8.36	1.67

where:

Ms_{ef} = the market share for end use e and fuel f

$Floor_{ief}$ = the square footage being used by customer i for end use e and fuel f

$Floor_i$ = the total square footage for customer i

W_i = the population weight for customer i.

Example 10-4: Estimating Market Share

Given the total square footage of buildings along with the square footage which is air conditioned, determine the market share for air conditioning. Assume all buildings use electric cooling equipment. The population weights for the strata are given in Example 10-3.

Building	Strata	Total Square Footage	Air Conditioning Square Footage
1	1	10,438	9,185
2	1	10,090	9,081
3	1	12,055	10,247
4	1	12,152	12,030
5	1	7,076	0
6	1	11,449	11,106
7	1	13,069	12,154
8	2	13,422	10,335
9	2	26,046	23,702
10	2	18,390	14,896
11	2	19,758	15,609
12	2	29,538	29,538
13	3	27,345	21,876
14	3	25,461	16,550
15	3	29,454	21,207

Step 1: Use Equation 10-5 to calculate the market share. The estimated market share for electric air conditioning is 84 percent. This was an easy example.

Demand Intensities

Demand intensities are similar in form to EUIs in that they both normalize a unit of energy by dividing by square footage. However, where the EUI contains annual energy, the demand intensity uses the peak end-use wattage divided by the square footage as shown in Equation 10-6. Alternatively, you could calculate the demand intensity for other fuels using BTUs instead of watts.

For consistency across buildings it's important that you use the estimated demand of the end uses on the same daytype and hour, such as "August weekday" at 6 P.M. It is typical for electric utilities to choose the system peak hour as the hour of interest since that is when the end uses have the greatest impact on the system.

$$DI_e = \frac{\sum_i (Watts_{ie} \cdot W_i)}{\sum_i (Floor_i \cdot W_i)}$$ (Equation 10-6)

where:

D_{ie} = demand intensity for end use e at a selected hour

$Watts_{ie}$ = wattage of end use e in building i at a selected hour

$Floor_i$ = total square footage of building i

W_i = population weight for building i.

Example 10-5: Estimating Demand Intensities

Given the data below, estimate the strata and population demand intensities for the major end uses shown. The population weights for each strata are also given on the following page.

Building	Strata	Square Footage	Peak End-Use Demand (kW)			
			Air Conditioning	Indoor Lights	Miscellaneous Equipment	Ventilation
1	1	10,438	5.3	4.7	4.3	3.6
2	1	10,090	8.0	5.2	4.7	2.4
3	1	12,055	7.2	6.1	6.2	3.9
4	1	12,152	7.1	6.2	5.3	4.4
5	1	7,076	5.8	3.8	4.7	2.6
6	1	11,449	5.8	4.7	5.9	2.8
7	1	13,069	10.7	5.1	8.6	4.8
8	2	13,422	11.0	7.2	6.9	4.6
9	2	26,046	14.3	13.8	13.0	8.5
10	2	18,390	11.2	10.4	10.5	5.7
11	2	19,758	13.6	12.4	12.4	5.7
12	2	29,538	17.4	11.9	17.4	6.8
13	3	27,345	18.7	13.6	13.7	9.0
14	3	25,461	20.1	10.5	13.0	6.2
15	3	29,454	14.8	16.5	17.9	10.4

Strata	Weight
1	27.4
2	21.4
3	8.4

Step 1: Calculate the strata level demand intensities by applying Equation 10-6 for those buildings within each strata. The results are shown in Table 10-3.

Table 10-3. Strata and population demand intensities.

Strata	Air Conditioning	Indoor Lights	Miscellaneous Equipment	Ventilation
1	0.65	0.47	0.52	0.32
2	0.63	0.52	0.56	0.29
3	0.65	0.49	0.54	0.31
Population	0.64	0.50	0.54	0.31

Step 2: Estimate the population demand intensities for each end use using Equation 10-6. See Table 10-3 for population

Coincidence Factors

Coincidence factors help describe the relationship between the peak demand for an end use within a building vs. the contribution that the end use has on system peak during the peak hour. We originally introduced coincidence factors in Chapter 2. Rewriting Equation 2-3 to solve for coincidence factor and take population weights into account, we obtain Equation 10-7.

$$CF_e = \frac{\sum_i Coin_{ie} \cdot W_i}{\sum_i Peak_{ie} \cdot W_i}$$ **(Equation 10-7)**

where:

CF_e = coincidence factor for end use e at given hour
$Coin_{ie}$ = load of end use e in building i at given hour
$Peak_{ie}$ = peak annual load of end use e in building i
W_i = population weight for building i.

Example 10-6: Estimating Coincidence Factors

Given the air conditioning annual peak load and load at 6 P.M. during the summer, calculate the strata and population coincidence factors for 6 P.M. on the August average weekday. The required data are presented below:

		Air Conditioning Load (kW)	
			Load @ 6 P.M.
Building	Strata	Peak Load	August Average Weekday
1	1	8.6	7.1
2	1	8.5	6.4
3	1	10.1	9.8
4	1	12.6	11.8
5	1	12.3	10.9
6	1	12.7	9.4
7	1	8.8	7.2
8	2	17.0	10.5
9	2	22.4	18.6
10	2	22.2	18.6
11	2	15.2	10.9
12	2	21.2	15.1
13	3	27.3	25.9
14	3	22.8	13.9
15	3	22.5	14.6
16	4	46.4	37.1

Strata	Weight
1	26
2	14
3	7
4	1

Step 1: Estimate the strata coincidence factors using Equation 10-7. The results are shown in Table 10-5.

Table 10-4. Estimated strata and population coincidence factors for air conditioning.

Strata	Coincidence Factor
1	0.85
2	0.75
3	0.75
4	0.95
Population	0.80

Step 2: Estimate the population coincidence factors using Equation 10-7. See Table 10-4 for results.

Full Load Hours

Full load hours are an interesting descriptor which can be obtained from end-use load shapes. Basically, they express the number of annual hours which an end use operates if it were running at full load. Of course, not all end uses operate at full load (e.g., motors) but many are fairly constant throughout the year (e.g., indoor lighting).

Full load hours are calculated by dividing the annual energy use for an end use by peak end use demand during the year as seen in Equation 10-7. You can also modify Equation 10-8 to accompany other fuels (e.g., for natural gas you would divide the annual gas consumption by the consumption during the peak hour).

$$FLH_e = \frac{\sum_i (kWh_{ie} \cdot W_i)}{\sum_i (kW_{ie} \cdot W_i)} \quad \textbf{(Equation 10-8)}$$

where:

FLH_e = the full load hours for end use e

kWh_{ie} = the annual energy use for building i and end use e

kW_{ie} = the annual peak demand for building i and end use e

W_i = the population weight for building i.

Example 10-7: Estimate End-Use Full-Load Hours

Given the annual peak demand, annual energy use, and population weights for the buildings shown in Table 10-5a, develop the strata and population estimates of full-load hours.

Strata	Weight
1	38.7
2	20.3
3	8.7
4	1.4

Step 1: Apply Equation 10-8 for both the strata and population estimates. The results are shown in Table 10-5b.

Table 10-5a. End-Use Data for Example 10-7.

		Annual End-Use Peak Demand (kW)					
Building	Strata	Air Conditioning	Space Heating	Indoor Lighting	Miscellaneous Equipment	Ventilation	Outdoor Lighting
1	1	4.95	1.27	2.57	2.54	1.71	1.00
2	1	3.68	0.96	2.77	3.08	2.37	0.94
3	1	3.05	1.36	3.27	3.21	2.17	0.86
4	1	5.03	1.48	3.77	2.99	1.49	0.61
5	1	4.20	1.53	2.81	3.50	1.57	1.03
6	1	4.24	1.64	2.97	3.63	2.23	0.84
7	1	3.96	1.19	3.34	3.53	2.31	1.02
8	2	8.63	3.41	7.88	6.61	5.15	1.56
9	2	10.14	2.53	7.95	5.01	3.30	1.22
10	2	9.90	3.01	6.55	6.36	3.82	1.24
11	2	8.08	2.69	7.21	7.51	3.50	1.22
12	2	8.32	1.85	5.88	6.16	3.50	1.68
13	3	9.74	3.17	8.72	9.82	6.75	2.19
14	3	9.03	4.16	7.11	9.24	6.87	3.29
15	3	11.52	4.32	13.03	7.32	5.61	2.55
16	4	19.48	5.76	12.42	9.24	7.24	2.79

(*cont'd*)

Table 10-5a. End-Use Data for Example 10-7 (cont'd).

	Annual End-Use Energy (kWh)					
Building	Air Conditioning	Space Heating	Indoor Lighting	Miscellaneous Equipment	Ventilation	Outdoor Lighting
1	6,681	1,636	11,950	15,007	6,595	3,603
2	5,098	1,054	13,228	19,698	14,576	3,359
3	3,214	1,554	11,095	17,206	10,745	2,494
4	6,253	1,542	13,108	21,672	7,222	1,376
5	6,117	1,354	10,005	18,973	7,937	2,806
6	4,217	1,616	15,552	19,015	12,785	2,510
7	4,548	1,666	18,189	25,586	11,558	3,675
8	10,831	4,927	41,592	38,248	31,942	4,935
9	9,004	2,692	39,297	23,802	19,092	4,217
10	10,549	2,933	34,297	46,099	19,910	3,523
11	11,097	3,013	21,746	43,914	20,066	4,217
12	7,881	2,528	29,558	47,276	21,343	3,986
13	14,185	3,160	28,127	74,169	37,292	4,490
14	9,088	4,147	30,677	50,090	40,461	8,480
15	14,185	4,113	64,408	35,669	30,701	7,320
16	22,372	7,741	41,622	48,402	39,622	8,581

Table 10-5b. Strata and population estimates of full load hours.

Building	Air Conditioning	Space Heating	Indoor Lighting	Miscellaneous Equipment	Ventilation	Outdoor Lighting
1	1,241	1,105	4,331	6,101	5,157	3,147
2	1,095	1,193	4,694	6,298	5,830	3,017
3	1,237	980	4,269	6,062	5,640	2,527
4	1,148	1,344	3,351	5,238	5,473	3,076
Population	1,182	1,122	4,418	6,091	5,366	2,906

A
Appendix: Assorted Data

Table A-1. Summer hourly end use and load research data (kW) for Example 8-4.

Site	Hour	Load Research Data	Air Conditioning	Indoor Lighting	Hot Water	Miscellaneous Equipment	Exterior Lighting	Engineering Total
1	1	2.42	0.00	0.25	0.00	0.30	2.00	2.55
1	2	2.14	0.00	0.25	0.00	0.30	2.00	2.55
1	3	2.04	0.00	0.25	0.00	0.30	2.00	2.55
1	4	2.04	0.00	0.25	0.00	0.30	2.00	2.55
1	5	2.45	0.00	0.25	0.00	0.30	2.00	2.55
1	6	2.52	0.00	0.25	0.00	0.30	2.00	2.55
1	7	1.94	0.00	1.00	0.00	1.00	0.00	2.00
1	8	4.50	0.00	3.00	0.00	2.00	0.00	5.00
1	9	6.72	0.00	5.00	0.00	3.00	0.00	8.00
1	10	7.04	0.00	5.00	0.00	3.00	0.00	8.00
1	11	7.84	0.00	5.00	0.00	3.00	0.00	8.00
1	12	7.04	0.00	5.00	0.00	3.00	0.00	8.00
1	13	7.68	0.00	5.00	0.00	3.00	0.00	8.00
1	14	7.36	0.00	5.00	0.00	3.00	0.00	8.00
1	15	7.28	0.00	5.00	0.00	3.00	0.00	8.00
1	16	7.68	0.00	5.00	0.00	3.00	0.00	8.00
1	17	6.64	0.00	5.00	0.00	3.00	0.00	8.00
1	18	5.94	0.00	3.00	0.00	3.00	0.00	6.00
1	19	2.91	0.00	1.00	0.00	2.00	0.00	3.00
1	20	1.11	0.00	0.25	0.00	1.00	0.00	1.25

1	21	0.49	0.00	0.25	0.00	0.30	0.00	0.55
1	22	2.42	0.00	0.25	0.00	0.30	2.00	2.55
1	23	2.40	0.00	0.25	0.00	0.30	2.00	2.55
1	24	2.35	0.00	0.25	0.00	0.30	2.00	2.55
2	1	4.94	0.00	0.10	0.05	1.00	4.00	5.15
2	2	4.79	0.00	0.10	0.05	1.00	4.00	5.15
2	3	4.12	0.00	0.10	0.05	1.00	4.00	5.15
2	4	4.58	0.00	0.10	0.05	1.00	4.00	5.15
2	5	4.48	0.00	0.10	0.05	1.00	4.00	5.15
2	6	4.12	0.00	0.10	0.05	1.00	4.00	5.15
2	7	7.79	0.00	2.00	0.20	2.00	4.00	8.20
2	8	13.43	0.00	6.00	0.60	4.00	4.00	14.60
2	9	13.65	0.20	8.00	0.80	6.00	0.00	15.00
2	10	15.29	0.60	8.00	1.00	6.00	0.00	15.60
2	11	13.12	1.00	8.00	1.00	6.00	0.00	16.00
2	12	13.20	1.50	8.00	1.00	6.00	0.00	16.50
2	13	17.00	2.00	8.00	1.00	6.00	0.00	17.00
2	14	15.75	2.50	8.00	1.00	6.00	0.00	17.50
2	15	16.20	3.00	8.00	1.00	6.00	0.00	18.00
2	16	15.75	2.50	8.00	1.00	6.00	0.00	17.50
2	17	13.40	1.50	6.00	0.60	6.00	0.00	14.10
2	18	6.24	0.50	3.00	0.20	4.00	0.00	7.70
2	19	6.74	0.20	1.00	0.05	2.00	4.00	7.25
2	20	4.52	0.05	0.10	0.05	1.00	4.00	5.20
2	21	5.00	0.00	0.10	0.05	1.00	4.00	5.15
2	22	4.94	0.00	0.10	0.05	1.00	4.00	5.15

Table A-1. Summer hourly end use and load research data (kW) for Example 8-4 (cont'd).

Site	Hour	Load Research Data	Air Conditioning	Indoor Lighting	Hot Water	Miscellaneous Equipment	Exterior Lighting	Engineering Total
2	23	4.17	0.00	0.10	0.05	1.00	4.00	5.15
2	24	4.94	0.00	0.10	0.05	1.00	4.00	5.15
3	1	13.77	1.00	2.00	0.00	2.00	12.00	17.00
3	2	13.60	1.00	2.00	0.00	2.00	12.00	17.00
3	3	14.96	1.00	2.00	0.00	2.00	12.00	17.00
3	4	14.79	1.00	2.00	0.00	2.00	12.00	17.00
3	5	14.28	1.00	2.00	0.00	2.00	12.00	17.00
3	6	17.00	1.00	2.00	0.00	2.00	12.00	17.00
3	7	15.30	1.00	5.00	0.00	4.00	8.00	18.00
3	8	20.37	3.00	8.00	0.00	6.00	4.00	21.00
3	9	21.39	5.00	9.00	0.00	8.00	1.00	23.00
3	10	22.62	7.00	10.00	0.00	8.00	1.00	26.00
3	11	25.76	9.00	10.00	0.00	8.00	1.00	28.00
3	12	30.38	12.00	9.00	0.00	9.00	1.00	31.00
3	13	31.85	14.00	10.00	0.00	10.00	1.00	35.00
3	14	36.00	15.00	11.00	0.00	9.00	1.00	36.00
3	15	31.50	16.00	10.00	0.00	8.00	1.00	35.00
3	16	29.05	17.00	9.00	0.00	8.00	1.00	35.00
3	17	30.72	18.00	7.00	0.00	6.00	1.00	32.00
3	18	24.57	17.00	5.00	0.00	4.00	1.00	27.00

3	19	18.27	15.00	3.00	0.00	2.00	1.00	21.00
3	20	16.60	13.00	1.00	0.00	2.00	4.00	20.00
3	21	18.70	11.00	1.00	0.00	2.00	8.00	22.00
3	22	14.25	8.00	1.00	0.00	2.00	4.00	15.00
3	23	13.05	4.00	1.00	0.00	2.00	8.00	15.00
3	24	10.56	1.00	1.00	0.00	2.00	8.00	12.00
4	1	6.02	0.00	4.00	0.00	3.00	0.00	7.00
4	2	5.82	0.00	3.00	0.00	3.00	0.00	6.00
4	3	4.65	0.00	2.00	0.00	3.00	0.00	5.00
4	4	3.88	0.00	1.00	0.00	3.00	0.00	4.00
4	5	3.72	0.00	1.00	0.00	3.00	0.00	4.00
4	6	3.84	0.00	1.00	0.00	3.00	0.00	4.00
4	7	4.90	0.00	2.00	0.00	3.00	0.00	5.00
4	8	8.30	0.00	6.00	0.00	4.00	0.00	10.00
4	9	14.85	0.00	10.00	0.00	5.00	0.00	15.00
4	10	13.80	0.00	10.00	0.00	5.00	0.00	15.00
4	11	14.55	0.00	10.00	0.00	5.00	0.00	15.00
4	12	12.30	0.00	10.00	0.00	5.00	0.00	15.00
4	13	14.55	0.00	10.00	0.00	5.00	0.00	15.00
4	14	13.44	0.00	11.00	0.00	5.00	0.00	16.00
4	15	13.50	0.00	10.00	0.00	5.00	0.00	15.00
4	16	12.30	0.00	10.00	0.00	5.00	0.00	15.00
4	17	14.10	0.00	10.00	0.00	5.00	0.00	15.00
4	18	14.25	0.00	10.00	0.00	5.00	0.00	15.00
4	19	13.95	0.00	10.00	0.00	5.00	0.00	15.00
4	20	11.05	0.00	9.00	0.00	4.00	0.00	13.00

Table A-1. Summer hourly end use and load research data (kW) for Example 8-4 (cont'd).

Site	Hour	Load Research Data	Air Conditioning	Indoor Lighting	Hot Water	Miscellaneous Equipment	Exterior Lighting	Engineering Total
4	21	9.13	0.00	8.00	0.00	3.00	0.00	11.00
4	22	8.20	0.00	7.00	0.00	3.00	0.00	10.00
4	23	7.65	0.00	6.00	0.00	3.00	0.00	9.00
4	24	6.40	0.00	5.00	0.00	3.00	0.00	8.00
5	1	7.59	0.00	0.15	0.08	0.00	6.00	6.23
5	2	7.10	0.00	0.15	0.08	0.00	6.00	6.23
5	3	6.54	0.00	0.15	0.08	0.00	6.00	6.23
5	4	8.34	0.00	0.15	0.08	0.00	6.00	6.23
5	5	8.15	0.00	0.15	0.08	0.00	6.00	6.23
5	6	6.72	0.00	0.15	0.08	0.00	6.00	6.23
5	7	12.00	0.00	3.00	0.30	0.00	6.00	9.30
5	8	15.58	0.00	9.00	0.90	0.00	6.00	15.90
5	9	19.98	0.30	12.00	1.20	0.00	0.00	13.50
5	10	15.84	0.90	12.00	1.50	0.00	0.00	14.40
5	11	20.10	1.50	12.00	1.50	0.00	0.00	15.00
5	12	16.54	2.25	12.00	1.50	0.00	0.00	15.75
5	13	15.68	3.00	12.00	1.50	0.00	0.00	16.50
5	14	15.53	3.75	12.00	1.50	0.00	0.00	17.25
5	15	20.70	4.50	12.00	1.50	0.00	0.00	18.00
5	16	15.87	3.75	12.00	1.50	0.00	0.00	17.25

Assorted Data **245**

5	17	18.10	2.25	9.00	0.90	0.00	0.00	12.15
5	18	6.05	0.75	4.50	0.30	0.00	0.00	5.55
5	19	11.26	0.30	1.50	0.08	0.00	6.00	7.88
5	20	8.82	0.08	0.15	0.08	0.00	6.00	6.30
5	21	7.10	0.00	0.15	0.08	0.00	6.00	6.23
5	22	7.59	0.00	0.15	0.08	0.00	6.00	6.23
5	23	6.54	0.00	0.15	0.08	0.00	6.00	6.23
5	24	5.91	0.00	0.15	0.08	0.00	6.00	6.23
6	1	21.93	1.50	3.00	0.00	3.00	18.00	25.50
6	2	25.25	1.50	3.00	0.00	3.00	18.00	25.50
6	3	23.72	1.50	3.00	0.00	3.00	18.00	25.50
6	4	21.93	1.50	3.00	0.00	3.00	18.00	25.50
6	5	23.97	1.50	3.00	0.00	3.00	18.00	25.50
6	6	22.19	1.50	3.00	0.00	3.00	18.00	25.50
6	7	26.19	1.50	7.50	0.00	3.00	18.00	25.50
6	8	26.15	1.50	3.00	0.00	6.00	12.00	27.00
6	9	27.95	4.50	12.00	0.00	9.00	6.00	31.50
6	10	38.61	7.50	13.50	0.00	12.00	1.50	34.50
6	11	38.64	10.50	15.00	0.00	12.00	1.50	39.00
6	12	44.64	13.50	15.00	0.00	12.00	1.50	42.00
6	13	45.68	18.00	13.50	0.00	13.50	1.50	46.50
6	14	52.92	21.00	15.00	0.00	15.00	1.50	52.50
6	15	45.68	22.50	16.50	0.00	13.50	1.50	54.00
6	16	48.30	24.00	15.00	0.00	12.00	1.50	52.50
6	17	42.72	25.50	13.50	0.00	12.00	1.50	52.50
6	18	34.83	27.00	10.50	0.00	9.00	1.50	48.00
6	18		25.50	7.50	0.00	6.00	1.50	40.50

Table A-1. Summer hourly end use and load research data (kW) for Example 8-4 (cont'd).

Site	Hour	Load Research Data	Air Conditioning	Indoor Lighting	Hot Water	Miscellaneous Equipment	Exterior Lighting	Engineering Total
6	19	30.24	22.50	4.50	0.00	3.00	1.50	31.50
6	20	30.00	19.50	1.50	0.00	3.00	6.00	30.00
6	21	30.36	16.50	1.50	0.00	3.00	12.00	33.00
6	22	22.05	12.00	1.50	0.00	3.00	6.00	22.50
6	23	19.80	6.00	1.50	0.00	3.00	12.00	22.50
6	24	17.64	1.50	1.50	0.00	3.00	12.00	18.00
7	1	9.45	0.00	6.00	0.00	4.50	0.00	10.50
7	2	7.47	0.00	4.50	0.00	4.50	0.00	9.00
7	3	6.00	0.00	3.00	0.00	4.50	0.00	7.50
7	4	5.04	0.00	1.50	0.00	4.50	0.00	6.00
7	5	5.82	0.00	1.50	0.00	4.50	0.00	6.00
7	6	5.46	0.00	1.50	0.00	4.50	0.00	6.00
7	7	6.08	0.00	3.00	0.00	4.50	0.00	7.50
7	8	15.00	0.00	9.00	0.00	6.00	0.00	15.00
7	9	19.13	0.00	15.00	0.00	7.50	0.00	22.50
7	10	22.28	0.00	15.00	0.00	7.50	0.00	22.50
7	11	21.83	0.00	15.00	0.00	7.50	0.00	22.50
7	12	21.83	0.00	15.00	0.00	7.50	0.00	22.50
7	13	19.35	0.00	15.00	0.00	7.50	0.00	22.50
7	14	20.64	0.00	16.50	0.00	7.50	0.00	24.00

7	15	21.38	0.00	15.00	0.00	7.50	0.00	22.50
7	16	20.70	0.00	15.00	0.00	7.50	0.00	22.50
7	17	21.83	0.00	15.00	0.00	7.50	0.00	22.50
7	18	20.70	0.00	15.00	0.00	7.50	0.00	22.50
7	19	22.50	0.00	15.00	0.00	7.50	0.00	22.50
7	20	16.19	0.00	13.50	0.00	6.00	0.00	19.50
7	21	14.36	0.00	12.00	0.00	4.50	0.00	16.50
7	22	12.15	0.00	10.50	0.00	4.50	0.00	15.00
7	23	10.80	0.00	9.00	0.00	4.50	0.00	13.50
7	24	9.60	0.00	7.50	0.00	4.50	0.00	12.00
8	1	11.47	0.00	0.23	0.11	2.25	9.00	11.59
8	2	10.08	0.00	0.23	0.11	2.25	9.00	11.59
8	3	10.08	0.00	0.23	0.11	2.25	9.00	11.59
8	4	9.97	0.00	0.23	0.11	2.25	9.00	11.59
8	5	10.66	0.00	0.23	0.11	2.25	9.00	11.59
8	6	11.24	0.00	0.23	0.11	2.25	9.00	11.59
8	7	17.16	0.00	4.50	0.45	4.50	9.00	18.45
8	8	27.27	0.00	13.50	1.35	9.00	9.00	32.85
8	9	33.41	0.45	18.00	1.80	13.50	0.00	33.75
8	10	34.40	1.35	18.00	2.25	13.50	0.00	35.10
8	11	30.96	2.25	18.00	2.25	13.50	0.00	36.00
8	12	34.90	3.38	18.00	2.25	13.50	0.00	37.13
8	13	38.25	4.50	18.00	2.25	13.50	0.00	38.25
8	14	36.62	5.63	18.00	2.25	13.50	0.00	39.38
8	15	40.50	6.75	18.00	2.25	13.50	0.00	40.50
8	16	35.04	5.63	18.00	2.25	13.50	0.00	39.38

Table A-1. Summer hourly end use and load research data (kW) for Example 8-4 (cont'd).

Site	Hour	Load Research Data	Air Conditioning	Indoor Lighting	Hot Water	Miscellaneous Equipment	Exterior Lighting	Engineering Total
8	17	30.77	3.38	13.50	1.35	13.50	0.00	31.73
8	18	16.11	1.13	6.75	0.45	9.00	0.00	17.33
8	19	15.33	0.45	2.25	0.11	4.50	9.00	16.31
8	20	9.71	0.11	0.23	0.11	2.25	9.00	11.70
8	21	11.12	0.00	0.23	0.11	2.25	9.00	11.59
8	22	11.59	0.00	0.23	0.11	2.25	9.00	11.59
8	23	11.47	0.00	0.23	0.11	2.25	9.00	11.59
8	24	9.62	0.00	0.23	0.11	2.25	9.00	11.59
9	1	33.66	2.25	4.50	0.00	4.50	27.00	38.25
9	2	32.13	2.25	4.50	0.00	4.50	27.00	38.25
9	3	35.96	2.25	4.50	0.00	4.50	27.00	38.25
9	4	35.57	2.25	4.50	0.00	4.50	27.00	38.25
9	5	32.51	2.25	4.50	0.00	4.50	27.00	38.25
9	6	30.60	2.25	4.50	0.00	4.50	27.00	38.25
9	7	40.10	2.25	11.25	0.00	9.00	18.00	40.50
9	8	42.53	6.75	18.00	0.00	13.50	9.00	47.25
9	9	49.16	11.25	20.25	0.00	18.00	2.25	51.75
9	10	49.14	15.75	22.50	0.00	18.00	2.25	58.50
9	11	60.48	20.25	22.50	0.00	18.00	2.25	63.00
9	12	68.36	27.00	20.25	0.00	20.25	2.25	69.75

9	13	65.36	31.50	22.50	0.00	22.50	2.25	78.75
9	14	70.47	33.75	24.75	0.00	20.25	2.25	81.00
9	15	72.45	36.00	22.50	0.00	18.00	2.25	78.75
9	16	68.51	38.25	20.25	0.00	18.00	2.25	78.75
9	17	62.64	40.50	15.75	0.00	13.50	2.25	72.00
9	18	55.28	38.25	11.25	0.00	9.00	2.25	60.75
9	19	43.00	33.75	6.75	0.00	4.50	2.25	47.25
9	20	45.00	29.25	2.25	0.00	4.50	9.00	45.00
9	21	40.59	24.75	2.25	0.00	4.50	18.00	49.50
9	22	30.71	18.00	2.25	0.00	4.50	9.00	33.75
9	23	31.05	9.00	2.25	0.00	4.50	18.00	33.75
9	24	22.14	2.25	2.25	0.00	4.50	18.00	27.00
10	1	38.89	0.00	9.00	0.00	6.75	30.00	45.75
10	2	40.02	0.00	6.75	0.00	6.75	30.00	43.50
10	3	41.25	0.00	4.50	0.00	6.75	30.00	41.25
10	4	37.44	0.00	2.25	0.00	6.75	30.00	39.00
10	5	32.76	0.00	2.25	0.00	6.75	30.00	39.00
10	6	35.88	0.00	2.25	0.00	6.75	30.00	39.00
10	7	21.79	0.00	4.50	0.00	6.75	15.00	26.25
10	8	20.48	0.00	13.50	0.00	9.00	0.00	22.50
10	9	31.39	0.00	22.50	0.00	11.25	0.00	33.75
10	10	27.34	0.00	22.50	0.00	11.25	0.00	33.75
10	11	27.00	0.00	22.50	0.00	11.25	0.00	33.75
10	12	28.35	0.00	22.50	0.00	11.25	0.00	33.75
10	13	33.75	0.00	22.50	0.00	11.25	0.00	33.75
10	14	32.76	0.00	24.75	0.00	11.25	0.00	36.00

Table A-1. Summer hourly end use and load research data (kW) for Example 8-4 (cont'd).

Site	Hour	Load Research Data	Air Conditioning	Indoor Lighting	Hot Water	Miscellaneous Equipment	Exterior Lighting	Engineering Total
10	15	28.01	0.00	22.50	0.00	11.25	0.00	33.75
10	16	30.71	0.00	22.50	0.00	11.25	0.00	33.75
10	17	27.68	0.00	22.50	0.00	11.25	0.00	33.75
10	18	33.75	0.00	22.50	0.00	11.25	0.00	33.75
10	19	31.39	0.00	22.50	0.00	11.25	0.00	33.75
10	20	44.25	0.00	20.25	0.00	9.00	15.00	44.25
10	21	47.63	0.00	18.00	0.00	6.75	30.00	54.75
10	22	48.83	0.00	15.75	0.00	6.75	30.00	52.50
10	23	42.21	0.00	13.50	0.00	6.75	30.00	50.25
10	24	41.28	0.00	11.25	0.00	6.75	30.00	48.00
11	1	30.49	0.00	0.34	0.17	3.38	30.00	33.88
11	2	30.83	0.00	0.34	0.17	3.38	30.00	33.88
11	3	27.44	0.00	0.34	0.17	3.38	30.00	33.88
11	4	29.48	0.00	0.34	0.17	3.38	30.00	33.88
11	5	30.49	0.00	0.34	0.17	3.38	30.00	33.88
11	6	31.85	0.00	0.34	0.17	3.38	30.00	33.88
11	7	27.13	0.00	6.75	0.68	6.75	15.00	29.18
11	8	47.80	0.00	20.25	2.03	13.50	13.50	49.28
11	9	43.54	0.68	27.00	2.70	20.25	0.00	50.63
11	10	47.91	2.03	27.00	3.38	20.25	0.00	52.65

11	11	51.84	3.38	27.00	3.38	20.25	0.00	54.00
11	12	45.66	5.06	27.00	3.38	20.25	0.00	55.69
11	13	52.79	6.75	27.00	3.38	20.25	0.00	57.38
11	14	50.20	8.44	27.00	3.38	20.25	0.00	59.06
11	15	49.82	10.13	27.00	3.38	20.25	0.00	60.75
11	16	59.06	8.44	27.00	3.38	20.25	0.00	59.06
11	17	43.30	5.06	20.25	2.03	20.25	0.00	47.59
11	18	22.35	1.69	10.13	0.68	13.50	0.00	25.99
11	19	23.49	0.68	3.38	0.17	6.75	13.50	24.47
11	20	18.10	0.17	0.34	0.17	3.38	15.00	19.05
11	21	28.46	0.00	0.34	0.17	3.38	30.00	33.88
11	22	33.20	0.00	0.34	0.17	3.38	30.00	33.88
11	23	30.49	0.00	0.34	0.17	3.38	30.00	33.88
11	24	33.88	0.00	0.34	0.17	3.38	30.00	33.88
12	1	38.44	3.38	6.75	0.00	6.75	30.00	46.88
12	2	40.78	3.38	6.75	0.00	6.75	30.00	46.88
12	3	41.25	3.38	6.75	0.00	6.75	30.00	46.88
12	4	41.72	3.38	6.75	0.00	6.75	30.00	46.88
12	5	42.19	3.38	6.75	0.00	6.75	30.00	46.88
12	6	42.19	3.38	6.75	0.00	6.75	30.00	46.88
12	7	39.00	3.38	16.88	0.00	13.50	15.00	48.75
12	8	59.54	10.13	27.00	0.00	20.25	13.50	70.88
12	9	76.85	16.88	30.38	0.00	27.00	3.38	77.63
12	10	73.71	23.63	33.75	0.00	27.00	3.38	87.75
12	11	93.56	30.38	33.75	0.00	27.00	3.38	94.50
12	12	96.26	40.50	30.38	0.00	30.38	3.38	104.63

Table A-1. Summer hourly end use and load research data (kW) for Example 8-4 *(cont'd)*.

Site	Hour	Load Research Data	Air Conditioning	Indoor Lighting	Hot Water	Miscellaneous Equipment	Exterior Lighting	Engineering Total
12	13	118.13	47.25	33.75	0.00	33.75	3.38	118.13
12	14	104.49	50.63	37.13	0.00	30.38	3.38	121.50
12	15	107.49	54.00	33.75	0.00	27.00	3.38	118.13
12	16	100.41	57.38	30.38	0.00	27.00	3.38	118.13
12	17	87.48	60.75	23.63	0.00	20.25	3.38	108.00
12	18	78.37	57.38	16.88	0.00	13.50	3.38	91.13
12	19	65.91	50.63	10.13	0.00	6.75	3.38	70.88
12	20	65.55	43.88	3.38	0.00	6.75	15.00	69.00
12	21	64.89	37.13	3.38	0.00	6.75	30.00	77.25
12	22	54.37	27.00	3.38	0.00	6.75	30.00	67.13
12	23	44.51	13.50	3.38	0.00	6.75	30.00	53.63
12	24	41.76	3.38	3.38	0.00	6.75	30.00	43.50
13	1	60.39	0.00	13.50	0.00	10.13	45.00	68.63
13	2	58.73	0.00	10.13	0.00	10.13	45.00	65.25
13	3	51.98	0.00	6.75	0.00	10.13	45.00	61.88
13	4	47.97	0.00	3.38	0.00	10.13	45.00	58.50
13	5	54.99	0.00	3.38	0.00	10.13	45.00	58.50
13	6	50.31	0.00	3.38	0.00	10.13	45.00	58.50
13	7	34.26	0.00	6.75	0.00	10.13	22.50	39.38
13	8	33.08	0.00	20.25	0.00	13.50	0.00	33.75

13	9	42.53	0.00	33.75	0.00	16.88	0.00	50.63
13	10	43.54	0.00	33.75	0.00	16.88	0.00	50.63
13	11	45.06	0.00	33.75	0.00	16.88	0.00	50.63
13	12	45.56	0.00	33.75	0.00	16.88	0.00	50.63
13	13	45.56	0.00	33.75	0.00	16.88	0.00	50.63
13	14	48.06	0.00	37.13	0.00	16.88	0.00	54.00
13	15	47.08	0.00	33.75	0.00	16.88	0.00	50.63
13	16	41.51	0.00	33.75	0.00	16.88	0.00	50.63
13	17	49.61	0.00	33.75	0.00	16.88	0.00	50.63
13	18	43.54	0.00	33.75	0.00	16.88	0.00	50.63
13	19	45.06	0.00	33.75	0.00	16.88	0.00	50.63
13	20	61.07	0.00	30.38	0.00	13.50	22.50	66.38
13	21	76.38	0.00	27.00	0.00	10.13	45.00	82.13
13	22	76.39	0.00	23.63	0.00	10.13	45.00	78.75
13	23	68.59	0.00	20.25	0.00	10.13	45.00	75.38
14	1	45.74	0.00	0.51	0.25	5.06	45.00	50.82
14	2	49.81	0.00	0.51	0.25	5.06	45.00	50.82
14	3	46.76	0.00	0.51	0.25	5.06	45.00	50.82
14	4	43.71	0.00	0.51	0.25	5.06	45.00	50.82
14	5	49.81	0.00	0.51	0.25	5.06	45.00	50.82
14	6	45.74	0.00	0.51	0.25	5.06	45.00	50.82
14	7	44.09	0.00	10.13	1.01	10.13	30.00	51.26
14	8	58.36	0.00	30.38	3.04	20.25	15.00	68.66
14	9	69.86	1.01	40.50	4.05	30.38	0.00	75.94
14	10	65.55	3.04	40.50	5.06	30.38	0.00	78.98
14	11	73.71	5.06	40.50	5.06	30.38	0.00	81.00

Assorted Data

Table A-1. Summer hourly end use and load research data (kW) for Example 8-4 (cont'd).

Site	Hour	Load Research Data	Air Conditioning	Indoor Lighting	Hot Water	Miscellaneous Equipment	Exterior Lighting	Engineering Total
14	12	77.68	7.59	40.50	5.06	30.38	0.00	83.53
14	13	81.76	10.13	40.50	5.06	30.38	0.00	86.06
14	14	71.76	12.66	40.50	5.06	30.38	0.00	88.59
14	15	81.10	15.19	40.50	5.06	30.38	0.00	91.13
14	16	78.85	12.66	40.50	5.06	30.38	0.00	88.59
14	17	65.67	7.59	30.38	3.04	30.38	0.00	71.38
14	18	31.19	2.53	15.19	1.01	20.25	0.00	38.98
14	19	25.48	1.01	5.06	0.25	10.13	15.00	31.45
14	20	30.30	0.25	0.51	0.25	5.06	30.00	36.08
14	21	41.67	0.00	0.51	0.25	5.06	45.00	50.82
14	22	47.26	0.00	0.51	0.25	5.06	45.00	50.82
14	23	45.23	0.00	0.51	0.25	5.06	45.00	50.82
14	24	70.56	0.00	16.88	0.00	10.13	45.00	72.00
14	24	41.17	0.00	0.51	0.25	5.06	45.00	50.82
15	1	61.17	5.06	10.13	0.00	10.13	45.00	70.31
15	2	61.88	5.06	10.13	0.00	10.13	45.00	70.31
15	3	69.61	5.06	10.13	0.00	10.13	45.00	70.31
15	4	61.88	5.06	10.13	0.00	10.13	45.00	70.31
15	5	56.25	5.06	10.13	0.00	10.13	45.00	70.31
15	6	60.47	5.06	10.13	0.00	10.13	45.00	70.31

15	7	51.79	5.06	25.31	0.00	20.25	5.06	55.69
15	8	79.28	15.19	40.50	0.00	30.38	5.06	91.13
15	9	93.15	25.31	45.56	0.00	40.50	5.06	116.44
15	10	123.73	35.44	50.63	0.00	40.50	5.06	131.63
15	11	131.83	45.56	50.63	0.00	40.50	5.06	141.75
15	12	131.83	60.75	45.56	0.00	45.56	5.06	156.94
15	13	152.38	70.88	50.63	0.00	50.63	5.06	177.19
15	14	151.27	75.94	55.69	0.00	45.56	5.06	182.25
15	15	157.70	81.00	50.63	0.00	40.50	5.06	177.19
15	16	150.61	86.06	45.56	0.00	40.50	5.06	177.19
15	17	158.76	91.13	35.44	0.00	30.38	5.06	162.00
15	18	121.65	86.06	25.31	0.00	20.25	5.06	136.69
15	19	97.81	75.94	15.19	0.00	10.13	5.06	106.31
15	20	79.18	65.81	5.06	0.00	10.13	5.06	86.06
15	21	108.92	55.69	5.06	0.00	10.13	45.00	115.88
15	22	99.68	40.50	5.06	0.00	10.13	45.00	100.69
15	23	64.35	20.25	5.06	0.00	10.13	45.00	80.44
15	24	52.20	5.06	5.06	0.00	10.13	45.00	65.25
16	1	89.56	0.00	20.25	0.00	15.19	67.50	102.94
16	2	84.17	0.00	15.19	0.00	15.19	67.50	97.88
16	3	82.60	0.00	10.13	0.00	15.19	67.50	92.81
16	4	81.61	0.00	5.06	0.00	15.19	67.50	87.75
16	5	76.34	0.00	5.06	0.00	15.19	67.50	87.75
16	6	71.08	0.00	5.06	0.00	15.19	67.50	87.75
16	7	53.16	0.00	10.13	0.00	15.19	33.75	59.06
16	8	50.63	0.00	30.38	0.00	20.25	0.00	50.63

Table A-1. Summer hourly end use and load research data (kW) for Example 8-4 *(cont'd)*.

Site	Hour	Load Research Data	Air Conditioning	Indoor Lighting	Hot Water	Miscellaneous Equipment	Exterior Lighting	Engineering Total
16	9	69.86	0.00	50.63	0.00	25.31	0.00	75.94
16	10	62.27	0.00	50.63	0.00	25.31	0.00	75.94
16	11	66.07	0.00	50.63	0.00	25.31	0.00	75.94
16	12	62.27	0.00	50.63	0.00	25.31	0.00	75.94
16	13	71.38	0.00	50.63	0.00	25.31	0.00	75.94
16	14	68.04	0.00	55.69	0.00	25.31	0.00	81.00
16	15	62.27	0.00	50.63	0.00	25.31	0.00	75.94
16	16	64.55	0.00	50.63	0.00	25.31	0.00	75.94
16	17	69.86	0.00	50.63	0.00	25.31	0.00	75.94
16	18	75.18	0.00	50.63	0.00	25.31	0.00	75.94
16	19	61.51	0.00	50.63	0.00	25.31	0.00	75.94
16	20	80.65	0.00	45.56	0.00	20.25	33.75	99.56
16	21	110.87	0.00	40.50	0.00	15.19	67.50	123.19
16	22	115.76	0.00	35.44	0.00	15.19	67.50	118.13
16	23	104.02	0.00	30.38	0.00	15.19	67.50	113.06
16	24	89.64	0.00	25.31	0.00	15.19	67.50	108.00
17	1	67.27	0.00	0.76	0.38	0.00	67.50	68.64
17	2	97.47	0.00	0.76	0.38	0.00	67.50	68.64
17	3	67.95	0.00	0.76	0.38	0.00	67.50	68.64
17	4	77.56	0.00	0.76	0.38	0.00	67.50	68.64

17	5	75.50	0.00	0.76	0.38	0.00	67.50	68.64
17	6	101.59	0.00	0.76	0.38	0.00	67.50	68.64
17	7	70.13	0.00	15.19	1.52	0.00	33.75	50.46
17	8	69.57	0.00	45.56	4.56	0.00	12.00	62.12
17	9	63.31	1.52	60.75	6.08	0.00	2.00	70.34
17	10	82.39	4.56	60.75	7.59	0.00	2.00	74.90
17	11	92.75	7.59	60.75	7.59	0.00	2.00	77.94
17	12	85.00	11.39	60.75	7.59	0.00	2.00	81.73
17	13	127.44	15.19	60.75	7.59	0.00	2.00	85.53
17	14	128.63	18.98	60.75	7.59	0.00	2.00	89.33
17	15	88.47	22.78	60.75	7.59	0.00	2.00	93.13
17	16	83.08	18.98	60.75	7.59	0.00	2.00	89.33
17	17	93.99	11.39	45.56	4.56	0.00	2.00	63.51
17	18	32.50	3.80	22.78	1.52	0.00	2.00	30.10
17	19	31.38	1.52	7.59	0.38	0.00	2.00	21.49
17	20	48.32	0.38	0.76	0.38	0.00	12.00	35.27
17	21	75.50	0.00	0.76	0.38	0.00	33.75	68.64
17	22	76.88	0.00	0.76	0.38	0.00	67.50	68.64
17	23	102.27	0.00	0.76	0.38	0.00	67.50	68.64
17	24	72.07	0.00	0.76	0.38	0.00	67.50	68.64
18	1	58.45	7.59	15.19	0.00	15.19	30.00	67.97
18	2	60.49	7.59	15.19	0.00	15.19	30.00	67.97
18	3	57.77	7.59	15.19	0.00	15.19	30.00	67.97
18	4	59.13	7.59	15.19	0.00	15.19	30.00	67.97
18	5	59.81	7.59	15.19	0.00	15.19	30.00	67.97
18	6	61.17	7.59	15.19	0.00	15.19	30.00	67.97

Table A-1. Summer hourly end use and load research data (kW) for Example 8-4 *(cont'd)*.

Site	Hour	Load Research Data	Air Conditioning	Indoor Lighting	Hot Water	Miscellaneous Equipment	Exterior Lighting	Engineering Total
18	7	89.12	7.59	37.97	0.00	30.38	15.00	90.94
18	8	137.09	22.78	60.75	0.00	45.56	8.00	137.09
18	9	172.91	37.97	68.34	0.00	60.75	7.59	174.66
18	10	193.49	53.16	75.94	0.00	60.75	7.59	197.44
18	11	189.24	68.34	75.94	0.00	60.75	7.59	212.63
18	12	202.45	91.13	68.34	0.00	68.34	7.59	235.41
18	13	249.83	106.31	75.94	0.00	75.94	7.59	265.78
18	14	229.64	113.91	83.53	0.00	68.34	7.59	273.38
18	15	233.89	121.50	75.94	0.00	60.75	7.59	265.78
18	16	244.52	129.09	68.34	0.00	60.75	7.59	265.78
18	17	216.27	136.69	53.16	0.00	45.56	7.59	243.00
18	18	188.63	129.09	37.97	0.00	30.38	7.59	205.03
18	19	154.68	113.91	22.78	0.00	15.19	7.59	159.47
18	20	114.66	98.72	7.59	0.00	15.19	15.00	136.50
18	21	117.23	83.53	7.59	0.00	15.19	30.00	136.31
18	22	94.23	60.75	7.59	0.00	15.19	30.00	113.53
18	23	76.50	30.38	7.59	0.00	15.19	30.00	83.16
18	24	52.53	7.59	7.59	0.00	15.19	30.00	60.38
19	1	74.84	0.00	30.38	0.00	22.78	30.00	83.16
19	2	62.72	0.00	22.78	0.00	22.78	30.00	75.56

Assorted Data

19	3	60.49	0.00	15.19	0.00	22.78	30.00	67.97
19	4	50.72	0.00	7.59	0.00	22.78	30.00	60.38
19	5	51.92	0.00	7.59	0.00	22.78	30.00	60.38
19	6	53.13	0.00	7.59	0.00	22.78	30.00	60.38
19	7	45.51	0.00	15.19	0.00	22.78	8.00	45.97
19	8	73.66	0.00	45.56	0.00	30.38	0.00	75.94
19	9	105.93	0.00	75.94	0.00	37.97	0.00	113.91
19	10	92.26	0.00	75.94	0.00	37.97	0.00	113.91
19	11	97.96	0.00	75.94	0.00	37.97	0.00	113.91
19	12	108.21	0.00	75.94	0.00	37.97	0.00	113.91
19	13	107.07	0.00	75.94	0.00	37.97	0.00	113.91
19	14	108.14	0.00	83.53	0.00	37.97	0.00	121.50
19	15	104.79	0.00	75.94	0.00	37.97	0.00	113.91
19	16	102.52	0.00	75.94	0.00	37.97	0.00	113.91
19	17	92.26	0.00	75.94	0.00	37.97	0.00	113.91
19	18	109.35	0.00	75.94	0.00	37.97	0.00	113.91
19	19	113.91	0.00	75.94	0.00	37.97	0.00	113.91
19	20	110.31	0.00	68.34	0.00	30.38	15.00	113.72
19	21	102.18	0.00	60.75	0.00	22.78	30.00	113.53
19	22	99.58	0.00	53.16	0.00	22.78	30.00	105.94
19	23	88.51	0.00	45.56	0.00	22.78	30.00	98.34
19	24	82.58	0.00	37.97	0.00	22.78	30.00	90.75
20	1	4.30	5.00	0.00	0.00	0.00	0.00	5.00
20	2	4.86	6.00	0.00	0.00	0.00	0.00	6.00
20	3	5.81	7.00	0.00	0.00	0.00	0.00	7.00
20	4	6.88	8.00	0.00	0.00	0.00	0.00	8.00

Table A-1. Summer hourly end use and load research data (kW) for Example 8-4 (cont'd).

Site	Hour	Load Research Data	Air Conditioning	Indoor Lighting	Hot Water	Miscellaneous Equipment	Exterior Lighting	Engineering Total
20	5	8.01	9.00	0.00	0.00	0.00	0.00	9.00
20	6	8.19	9.00	0.00	0.00	0.00	0.00	9.00
20	7	9.10	10.00	0.00	0.00	0.00	0.00	10.00
20	8	8.90	10.00	0.00	0.00	0.00	0.00	10.00
20	9	9.90	11.00	0.00	0.00	0.00	0.00	11.00
20	10	10.12	11.00	0.00	0.00	0.00	0.00	11.00
20	11	9.72	12.00	0.00	0.00	0.00	0.00	12.00
20	12	16.74	17.09	0.00	0.00	0.00	0.00	17.09
20	13	20.50	22.78	0.00	0.00	0.00	0.00	22.78
20	14	23.64	28.48	0.00	0.00	0.00	0.00	28.48
20	15	28.02	34.17	0.00	0.00	0.00	0.00	34.17
20	16	25.34	28.48	0.00	0.00	0.00	0.00	28.48
20	17	14.69	17.09	0.00	0.00	0.00	0.00	17.09
20	18	12.60	15.00	0.00	0.00	0.00	0.00	15.00
20	19	12.35	13.00	0.00	0.00	0.00	0.00	13.00
20	20	8.90	10.00	0.00	0.00	0.00	0.00	10.00
20	21	6.23	7.00	0.00	0.00	0.00	0.00	7.00
20	22	3.80	4.00	0.00	0.00	0.00	0.00	4.00
20	23	2.73	3.00	0.00	0.00	0.00	0.00	3.00
20	24	0.98	1.00	0.00	0.00	0.00	0.00	1.00

Index

actual load, 12, 14
aggregation, 113–127, 162, 199
air conditioning, 161, 186–187, 209–210
allocation
   Neyman allocation, 48–50
   proportional allocation, 50–51
appliance saturation, 13
building simulation, 129–130
calibration
   annual energy, 162–163
   monthly energy, 163
   hourly load, 163–164
capacity, 7
CDA, 9, 167
   annual, 167–174
   hourly, 185, 193–194
   monthly, 174–179
   time-series, 180–184
class, 1–3
cleanness of metered data, 109–112
coefficient of variation, 34
coincidence factor, 13–14, 231–233
coincident load, 14
condensing data, 112–113
confidence level, 33–34
connected load, 11–12
cooking
   four-mode model, 150–152
   two-mode model, 152–153
   cooking intensity model, 154–156
   CDA, 190
customer tracking, 98–100
data collection, 7–8
   customer tracking, 98–100
   mail surveys, 78–81
   on-site, 82–97
   phone surveys, 81–82
   survey design, 77–78
daylight sensors, 135–137
daytypes, 18–22, 112–113
Delenius-Hodges stratification, 6, 46–48
demand
   average, 54–55
   instantaneous, 54
   intensities, 229–231
   moving average, 54–55
   savings, 14–15
demand-side-management, 3–5, 14–15, 172
diversity factor, 13–14

DOE-2, 24, 129, 160
domestic hot water
  age group model, 148
  CDA, 189–190
  consumption model, 149–150
  linear model, 146
  square root model, 147
  standby losses, 148–149
electrical demand, 53–56
end-use
  categories, 15
  load shapes, 1, 5, 8–10
  metering, 23, 53–76, 106–109
end-uses
  air-conditioning, 161, 186–187, 209–210
  cooking, 150–157, 190
  domestic hot water, 145–150, 189–190
  exterior lighting, 137–141, 191–192
  interior lighting, 132–137, 188–189
  miscellaneous, 141–145, 192
  refrigeration, 156–159, 190–191
  space heating, 160, 209–210
  ventilation, 159, 187–188
energy savings, 14–15
energy usage, 53–56
estimates
  coincidence factor, 13–14, 231–233
  demand intensities, 229–231
  EUI, 224–225
  floorspace, 221–224
  full load hours, 13, 233–237
  market share, 225–229
EUI, 94
full load hours, 13, 233–237
floorspace, 221–224
HSEM, 167, 198–199

industrial process
  box model, 141–142
  on-off model, 142–143
  sawtooth model, 143–144
  sinusoidal model, 144–145
interior lighting models
  box model, 132
  CDA, 185–189
  space model, 132–133
  occupancy sensor model, 133–135
load
  actual, 12, 14
  capacity, 7
  coincident, 14
  connected, 11–12
  constant with fixed operating hours, 58–59
  constant with variable hours, 59–61
load research, 102
load shapes, 1–2, 8–10
  aggregation, 113–127, 162, 199
  average, 125–127
  calibration, 162–165
  development, 5, 22–23
  engineering models, 23–24
  metered data, 23
  methods, 22–27
  statistical analysis, 24–26
  statistical/engineering models, 27, 167, 198–199
  transfer, 201–220
  typical, 119–125
mail surveys
  accuracy, 78–79
  premise definition, 81
  response rate, 78
  self-contained, 80–81
market share, 13, 225–229

mean, 34–35
meter types, 56–57
metering
  costs, 57–58
  data certification, 73–76
  equipment, 6–7
  installation, 69–73
  plans 61–62, 66–69
  range checks, 73–75
  schedule checks, 75–76
miscellaneous equipment
  box model, 141–142
  CDA, 192
  on-off model, 142–143
  sawtooth model, 143–144
  sinusoidal model, 144–145
nameplate capacity, 12
Neyman allocation, 6, 48–50
on-site data collection
  accuracy, 83
  customer tracking, 98–100
  quality control, 94–97
  response rate, 82–83
  scheduling, 84–86
  touring facilities, 93–94
  training auditors, 86–93
part load factor, 13
peak hour, 11–12
phone surveys
  accuracy, 82
  response rate, 81–82
population
  estimating, 29–30
  size, 32–33
  weight, 9–10
precision, 32–33
premise definition, 101–102
  multiple meters-multiple
    businesses, 105–106
  multiple meters-single
    business, 103–105

single meter-multiple
  businesses, 103
single meter-single business,
  102–103
refrigeration/freezers, 156–157
  case types, 157
  CDA, 190–191
  demand, 158
  energy use, 157
  refrigerating effect, 158
residential classifications, 18
response rate, 78, 81–83
SAE, 167, 194–198
sample
  design, 6, 29–51
  frame, 30–31
  target variable, 31s–32
sampling techniques
  Delenius-Hodges stratification,
    46–48
  Neyman allocation, 48–50
  proportional allocation, 50–51
  random sampling, 35–38
  stratification by building
    type, 40–46
  stratification within building
    type, 46–51
  stratified sampling, 38–51
  uniform stratification, 29–40
SIC, 16–17
solar
  apparent solar time, 138–139
  hour angle of sunrise, 137
  sunrise/sunset time, 139
space heating, 160, 185–186,
  209–210
standard deviation, 34–35
statistics
  CDA, 9, 167 185
  coefficient of variation, 34
  confidence level, 34–35

Delenius-Hodges stratification, 6
Neyman allocation, 6
mean, 34–35
precision, 32–33
sample design, 6, 29–51
standard deviation, 34–35
standard dispersion, 33
standard error, 3
stratification
  by building type, 40–46
  Delenius-Hodges stratification, 46–48
  uniform, 29–40
survey design, 77–78
system peak hour, 11–12
terminology, 11–27
transfer
  class adjustment, 201, 208–219
  class buildup, 201–207
  hybrid, 201, 219–220
ventilation, 159, 187–188
whole-building loads, 3, 18–19, 102
z-value, 33–34